INFORMATION BEHAVIOUR OF SCIENTISTS AND ENGINEERS

CONCEPTS IN COMMUNICATION INFORMATICS & LIBRARIANSHIP (CICIL)

For a list of titles in the series, see pp. 385-388 at the end of the book

---

About the Author

Dr. M.S. Sridhar is a post-graduate in mathematics and business management and a doctorate in library and information science. He is in the profession for last two decades. For last 16 years he is heading the library and documentation division of ISRO Satellite Centre, Bangalore. Earlier he has worked in the libraries of National Aeronautical Laboratory (Bangalore), Indian Institute of Management (Bangalore) and University of Mysore. Dr. Sridhar has published about 35 research papers in professional journals, presented a number of papers in conferences and seminars, delivered several professional lectures and published two books entitled *User Research : A Review of Information Behaviour Studies in Science and Technology* and *Problems of Collection Developments in Special Libraries.*

Concepts in Communication
Informatics & Librarianship—68

General Editor : **S.P. Agrawal**

INFORMATION BEHAVIOUR OF SCIENTISTS AND ENGINEERS

A Case Study of Indian Space Technologists

M.S. SRIDHAR

Head, Library and Documentation
ISRO Satellite Centre
Bangalore

CONCEPT PUBLISHING COMPANY, NEW DELHI-110059

ISBN 81-7022-563-9

First Published 1995

Published and Printed by
Ashok Kumar Mittal
Concept Publishing Company
A/15-16, Commercial Block, Mohan Garden
NEW DELHI-110059 (India)

DEDICATED TO

MY MOTHER SMT. RAMAMMA

CONTENTS

GENERAL EDITOR'S INTRODUCTION

Information is the life - blood of the researcher made in any field of activity or discipline. Particularly in scientific and technological researcher scientific communication is 'a salient feature of a scientific product'. Of late there is a growing realisation that scientists should be spared from the labour involved in gathering of their data and the time and energy so saved should be utilised by them for their research work. But this is possible only when someone well-versed in the technique of information collection does this job for the scientists. This someone can be no other person than the librarian or the informatin scientist who is trained for the purpose and whose function is to feed the scientists with their required information. But the librarians/information scientists cannot be an effective producer of information if they are ignorant about information behaviour and characteristics of the consumer of their information. It is not enóugh that they have a thorough grounding in information retrieval method and has technical competency. It is very essential for them that they should have a full rapport with their clientele. In other words they should be aware of not only the information behaviour of their users, but also should know their characteristics *i.e.*, their habit, their style of work, their preferences, etc., Information retrieval system for the benefit of information seeker must reflect this human factor. Unfortunately most of the present day librarian and the information scientists overlook this factor in formulation of the system. They believe that their function is restricted to only knowing the new methods of recording and desseminating information. They do not bother to understand the attitude and information —gathering pattern of their users. "A sufficient identification, analysis and co-ordination of the 'real' information — needs of the user is an essential basis for the planning, implementation and operation of information — systems and network".

In India the research output in library and information science is scanty. Moreover, much of the research which the professional has produced is not related to professional practices which may help improve the library and the information service. It is a matter of great concern that the number of user — studies done by Indian library and information scientists is very small. Of course, there are user - studies pertaining to scientists. But we have a very few studies on technologists. It is desirable that the librarians and information scientists should devote more attention towards this neglected field.

The present study, which attempts to investigate the information — transfer process and determine the information seeking behaviour of Indian space technologists of ISRO Satellite Centre, Bangalore and which also attempts to find out the user — Characteristics of these technologists and Co-ordinate their information seeking behaviour with their characteristics, is a pioneer work indeed in the field of space technology in India. The study is significant in many ways. It shows how the behavioural pattern of the space technologists is different from that of the scientists. The space technologists primarily seek information for their self-improvement and for keeping themselves up-to-date in their area of specialisation with a view to improving their professional efficiency and getting self-satisfaction with their work. Unlike scientists they seem to be unconcerned with their recognition. Nor they seek information for competing with their peers or for writing or publishing for their career improvement. The study makes an extensive comparison of its findings with the findings of the studies of the similar nature in developed countries. It may also be mentioned that the study makes a suggestion for setting up a multi-disciplinary research organisation at national level on the pattern of similar organisations established for the purpose in western countries. The main objective of this national organisation will be to overview the researchers made in communication and information behaviour of the Indian scientists, engineers, technicians and technologists.

S.P. AGRAWAL

PREFACE

Though libraries exist basically for users, research in Indian librarianship has taken for granted the user component of the system by importing findings of user research done abroad. As a student of M.B.A., I was very much convinced that there is a greater need for customer focus and extensive and indepth customer related studies in librarianship. The same has landed me later in initiating (1982) and successfully completing (1987) a doctoral research work on information seeking behaviour (ISB) of the Indian Space Technologists (IST) of ISRO Satellite Centre (ISAC), Bangalore. Couple of professional colleagues have evinced interest in making known this study to wider professional community. So born this book out of my doctoral work.

Review of relevant literature is excluded from this book as my other book entitled *User-research: A Review of Information Behaviour Studies in Science and Technology* (Sridhar, 1990) provides more detailed and comprehensive review of literature. However, wherever necessary other related works are extensively quoted and cited in the book.

I remain grateful to my guide and well wisher Dr. H.A. Khan, Professor, Department of Library and Information Science, University of Mysore for his guidance and help. I am indebted to authorities, scientists, engineers, technicians and colleagues at ISAC, the University of Mysore and many others who have helped and encouraged me in this research work. My thanks are also due to Concept Publishing Company for publishing the book.

Bangalore **M.S. SRIDHAR**

LIST OF TABLES

LIST OF FIGURES

SYMBOLS, ABBREVIATIONS AND ACRONYMS

χ^2	-	Chi-square Statistic
AAES	-	American Association of Engineering Societies
ACSP	-	Advisory Council of Scientific Policy
ADI	-	American Documentation Institute
AERE	-	Atomic Energy Research Establishment (UK)
AFIT	-	Air Force Institute of Technology
AFSC	-	Air Force Speciality Code
AGARD	-	Advisory Group for Aeronautical Research and Development(NATO)
AIAA	-	American Institute of Aeronautics and Astronautics
AIP	-	American Institute of Physics
APA	-	American Psychological Association
APPLE	-	Ariane Passenger Payload Experiment
ARIST	-	Annual Review of Information Science and Technology
ASIS	-	American Society for Information Science
ASLIB	-	Association of Special Libraries and Information Bureau
BASR	-	Bureau of Applied Social Research (Colunbia University)
BLLD	-	British Library Lending Division
BLR & DD	-	British Library Research & Development Division
CBRI	-	Central Building Research Institute
CRC	-	Centre for Research on Communications (of JHU)
CRS	-	Congressional Research Service
CRUS	-	Centre for Research on User Studies (University of Shiffield)
CRUSK	-	Centre for Research on Utilization of Scientific Knowledge

D&U	-	Dissemination and Utilization
DDC	-	Defence Documentation Centre (USA)
df	-	Degrees of Freedom
DOD	-	Department of Defence (USA)
DOE	-	Department of Energy (USA)
DOS	-	Department of Space (India)
DPC	-	Departmental Promotion Committee
EJC	-	Engineering Joint Council (USA)
ERIC	-	Educational Research Information Centre (USA)
ESA	-	European Space Agency
GSFC	-	Goddard Space Flight Centre (NASA)
HIP	-	High Information Potential
IASLIC	-	Indian Association of Special Libraries & Information Centre
IAU	-	International Astronomical Union
ICR	-	Institute for Communication Research (Stanford Univeristy)
IEE	-	Institution of Electrical Engineers (UK)
IEEE	-	Institute of Electrical and Electronics Engineers (USA)
IEG	-	Information Exchange Group (NIH Project)
IISc	-	Indian Institute of Science
IME	-	Institution of Mechanical Engineers
INFROSS	-	Information Requirements of the Social Sciences
INSAT	-	Indian National Satellite
IRS	-	Indian Remote Sensing Satellite
ISAC	-	ISRO Satellite Centre
ISB	-	Information Seeking Behaviour
ISI	-	Indian Standards Institution
ISR	-	Institute for Social Research (University of Michigan)
ISRO	-	Indian Space Research Organisation
ISSP	-	Indian Soviet Satellite Project (Indian Scientific

		Satellite Project)
IST	-	Indian Space Technologists
JHU	-	Johns Hopkins University
JSS	-	Joint Services Specification (India, Ministry of Defence)
LC	-	Library of Congress (USA)
LIP	-	Low Information Potential
MEDLARS	-	Medical Literature Analysis and Retrieval System
MIL	-	Military (US Military standards)
MIT	-	Massachussets Institute of Technology
NAL	-	National Aeronautical Laboratory
NASA	-	National Aeronautics and Space Administration
NIH	-	National Institute of Health
NISP	-	National Information System for Psychology (APA Project)
NISPA	-	National Information System for Physics and Astronomy (AIP Project)
NISSAT	-	National Information System for Science and Technology (India)
NLL	-	National Lending Library (Later BLLD)
NSF	-	National Science Foundation
NTIS	-	National Technical Information Service
ORG	-	Operations Research Group (Case Institute of Technology)
p	-	Statistical Significance Level
PIE	-	Physics Information exchange (AIP Project)
r	-	Pearson Product Moment Correlation Coefficient
r_s	-	Spearman Rank Order Correlation Coefficient
RAC	-	Reliability Analysis Centre
RAE	-	Royal Aeronautical Establishment (UK)
RAM	-	Random Alarm Mechanism

RDT & E	-	Research, Development, Test and Evaluation
SD	-	Standard Deviation (SD)
SDR	-	Solution Development Record
SIE	-	Scientific Information Exchange (APA Project)
SROSS	-	Stretched Rohini Satellite Series
SHAR	-	Shriharikota Range (ISRO Centre)
SIRED	-	Survey of Information Requirements of Engineering Designers (UK, Ministry of Technology)
STI	-	Scientific and Technical Information
t	-	t-Test Value
TDP	-	Technology Development Programme
TEA Lasers	-	Transversely Excited Atmospheric Pressure CO_2 Lasers
TTC	-	Telemetry, Tracking and Communications
UDC	-	Universal Decimal Classification
VITM	-	Vishveshwaraya Industrial and Technological Museum
VSSC	-	Vikram Sarabhai Space Centre
WFEO	-	World Federation of Engineering Organisations.

1

INTRODUCTION

1.1 Introduction

Information is an amorphous concept, less susceptible to a precise definition. Yet, everyone has to deal with it in many ways throughout one's life. "Indeed, information has been described as the fifth need of man ranking after air, water, food and shelter" (Kemp, 1976, p.101). Information-collection, transfer and use are all - pervasive and universal activities in all walks of life. It was conservatively estimated in 1972 in USA ". . . that R & D organisations invest 1.25 million dollars in a 30 year career of a scientist or engineer" (Vincent and Mirakhor, 1972, p. 45) and ". . . upto 90 per cent of the work involved in any white-collar job involves the seeking and obtaining of information" (Evans, 1976, p. 266). This enables one to see that an enormous amount of money is indirectly being spent on information-related activities of any work. A study of U.S. industries in 1967 revealed that the cost of acquiring information from outside-the-firm (excluding overhead) ranged from $ 2170 to $ 7550 per professional employee per annum (Gilmore *et al.*, 1967, p.113). Thus, knowingly or unknowingly, intentionally or unintentionally, all of us most of the time of our life and work (including the actions of the author and the reader at this moment) are concerned with information — its generation, recording, processing, repackaging, transfer, receiving, use and application. In the words of Debons (1974, p xiii), "information enfolds man physically and spiritually. It is the vehicle and substance ... [and] . . . is a resource . . . for ... power . . .".

Knowledge[1] in general and information in particular become more meaningful when they are transferred and communicated. The concepts 'information' and 'communication' are so coexistent that they are often used synonymously without much distinction in common usage and one often implies the other in many situations. In the context of scientific communication, Garvey (1979, p. 2) while emphasising the fact that communication is the essence of science says, ". . . communicability becomes a salient feature of a scientific product. . .". Further, information is very closely related to communication. A discussion of the one brings to the fore, the other. The information 'explodes' into power only when it is transferred and communicated. In other words, information is activised by communication.

Information and communication are equally important in the efficient and effective management of any enterprise. They have definite relations with performance of R & D organisations and industries as well. Based on a series of empirical studies Rothwell and Robertson (1975, p.393) concluded that there exists ". . . a positive relationship between good and efficient communication and successful industrial innovation". Weinberg (1967) emphasises the same in a slightly different way when he says ". . . though it is true that poor management can and does occur with the best of communication systems, poor communication almost always leads to bad management".

The purpose of transfer and communication of information is 'use'. Without the intention on the part of the ultimate receiver or beneficiary of information to use it, the whole exercise of transfer and communication becomes futile and information transferred becomes redundant. On the other hand, seeking and communicating information are two sides of the same coin from the viewpoint of individuals concerned. Both have the same purpose : use. Information which has no use is no information. Thus the central thread of the whole range of activities relating to information-transfer and communication is 'use'. Hence the emphasis on 'use' and user-orientation to communication and information — whether from a formally designed and operated information-system or from an informal source. A wide range of research works centered around 'use' and user of information called 'use and user-studies' have cropped up in the last four decades.

1.2 The Need for the Study

Like science, technology is also a human enterprise and "in all human activities, it is not ideas or machines that dominate; it is people" (Hills, 1982, p.11). Space technology itself is concerned with 'information transactions' (Dhawan, 1983, p.7) and communication and space technologists who work toward this goal through their experiments, research and development form a specific category of users of information.

The user is the key person in any information - system. All the luxuries of information-revolution and problems of information-explosion are centered around the user and his convenience. Understanding the user is half the battle in providing information-services. The success of any information-system depends considerably on how best the system design is based on a close and accurate understanding of the users. The user is not only the most important aspect, but is also, paradoxically, a dynamic component of information-system. As such, understanding 'user' is an important and a continuous activity. 'Know thy customer' is the cardinal rule of any business enterprise. Accordingly, extensive market research, customer-behaviour studies and demand forecasting are carried out in business. A formal information centre or library is yet to pick up these activities in the same spirit. It may be partly due to not-for-profit and paternalistic nature of information-systems and partly due to practical difficulties connected with the factors such as uniqueness of information-services, dual role of information as raw material as well as product or service, dual role of customer as producer and consumer of information, the competing and complimentary informal system of communication. Infact, information-system is not concerned with just demand, it has to stretch its hands to know the information-needs, the motives and purposes of seeking information, ways and means of gathering information and the entire user-attitudes and practices in relation to information.

There appears to be a lopsided emphasis in library and information field on user-education than on understanding the user. This is largely due to the systems designed and implemented, which are unfriendly, complex and not concerned with the perceived needs of users. What is more important is intensive librarian-education about users than mere education of users of the existing system (Thompson, 1982, p.11). Hoadley and Clark (1972, p.133) say that a library can achieve its goal "... if the library is more precise about who its users are. This precision, coupled with more research into behaviour and information-gathering patterns of these user groups, will assist the library more effectively in developing programmes and using its resources and limited funds to achieve desired goals." Accurate and up-to-date knowledge about users and their information-behaviour is one of the essential ingredients for designing an information-system.

The need is for user-driven design than technology-driven design and more research on human variables of information-system than technology and system variables (Atherton, 1975, p.672). Unfortunately, the emphasis in the profession has been on imparting technical skills than information skills.

The effectiveness of library and information-system depends on the extent to which the system-characteristics correspond with the user and on how much the potential user is willing and able to make use of it. Kunz and others (1977, p.9) say "A sufficient identification, analysis and co-ordination of the 'real' information-needs of the user is an essential basis for the planning, implementation and operation of information-systems and networks." In addition to "this 'objective aspect', the 'subjective aspect' of information-system—*i.e.*, in the eyes of [the user] . . . is unjustly neglected. Disregarded is the fact that a particular class of people will accept modes of information-transfer only if they are applicable to their habits, their style of work, etc., and only if certain minimum practical requirements are met".

System designers, planners and managers have, by and large, failed to properly consider the role of human factors and their effect on acceptance and utilisation of information (Mick, 1980, p.21). Oldman (1976, p.1) reviewing the research in librarianship says "A particular style of systems analysis has dominated library management research both in this country (U.K.) and in the United States in the last decade. The starting point for this research has been the library system. A more important starting point is the *target* for information —the individuals who constitute the 'receiving' community." "It is well recognised that the user has been given inadequate consideration from almost any point of view in most information—supplying systems existing today . . ." Landau and others (1975, p.422) further say ". . . in a great majority of cases, the user is neither understood nor studied nor, in some cases, even identified until after the system is already evolved or in a completely operational state. Some have characterized this as an 'upside-down system design."[2]

Research in library and information science is more meaningful if the findings increase the efficiency and effectiveness of information flow. To do this the emphasis should shift from system or service-oriented research to user-oriented research. Like in business, efforts should be put in primarily to relate product or system design to the perceived needs of those for whom the product or system is intended and the system operation should be guided by the knowledge about the user (Cronin, 1981, p.37).

Finally, selling or promoting the information-services and user-education should only take a secondary position subject to the condition that the system design is based on a proper understanding of the user. In this way only a librarian can improve upon his role as an information-transfer agent. He has to intimately and individually understand the requirements of his users and continuously update his knowledge about users through systematic studies and observations.

There is another very interesting line of argument from Garvey about why librarians should actively involve themselves in user-related research, communication and information-transfer process. Himself a scientist, Garvey (1979, p.5) analyses the role of librarians and pleads "the sooner librarians can interact with scientists in this process [of communication] the better they can fulfil their information-service functions." His argument is that "a real scientist cannot become sole 'information man' unless he sacrifices his research productivity" (p.10) and "since it is not likely that the typical scientist is going to become an expert in librarianship, then the librarian must become expert in the communication structure of the world of science" (p.16). In the words of Foskett (1984, p.61) "it is not enough for librarians merely to respond to new ways of recording and disseminating information." A more creative approach will come from a much greater involvement with both producers and consumers of information.

User-studies are necessary not only for planning and designing information systems but also for their efficient and effective operation. "A systematic and comparative study often reveals unanticipated data, which may prompt hitherto unconsidered courses of action" (Rowley and Turner, 1978, p.56). There will be many surprises to even an experienced librarian about his own clientele as he would not have heard them all in a systematically planned setup in his day-to-day work. The obvious drawback in relying on experience is bias towards outspoken and intensive users, which is, at times, either too pleasant or too bitter.

Incidentally, user-studies would help to improve the relation of the library with users in an objective way provided it is done within the organisation concerned. An organisation-oriented user-study like the present one helps to stratify users into some homogeneous groups by various characteristics of users, their information- seeking behaviour and their interaction with the information system. Many similarities and dissimilarities among user groups would help to increase the effectiveness of information-services and information-transfer process by focusing information-services to appropriate groups.

More user-studies are needed not just to determine and confirm the general patterns of information-gathering, but to identify departures from the norm which, even if only in small areas, have significant practical effects (Hanson, 1974). The study of communication-behaviour of users within the population would help librarians to identify and understand linkage roles of users and how their informal contacts complement and supplement the formal sources of information. Finally, as mentioned earlier, user-studies help to check and improve upon the place of existing information system by objective performance evaluation and to formulate future policies.

We have seen the importance and the general lack of the user-related research. For various reasons, the literature on user-research has necessarily been voluminous. Yet, findings are disproportionately smaller. Ignoring extensive data, repetitive as well as fragmented presentations, the literature does not appear to be so voluminous. The behaviour-related research in information-storage and retrieval is not great — either absolutely or relatively. Very little is known about user's preferences which can be used with confidence in the design of a new information-retrieval system. Due to its contingency nature very few studies can be considered to be relevant for a given user-population. The means for gathering needed information may be similar all over the world, but styles do differ according to local conditions (Lingwood and McAnany, 1971, p.16). As such there is a need for localised comprehensive studies looking for correlation between various factors before they are generalised.

Markee (1972, p.119) stresses the need for user-research in the parent organisation and says that "a review of the library literature reveals that the amount of library research being done by the librarian at his institution is not in proportion to his professional responsibilities." He answers the question as to where the research should originate by saying that the "library research should begin and continue right in the parent institution." This bottom-upward approach can only provide a solid foundation for national information systems of developed and developing countries.

Researchers in the past have repeatedly pointed out that unusually more attention has been concentrated on information-behaviour of scientists than that of engineers, technologists, technicians and technocrats. (Passman, 1969, p.8; Fishenden, 1965, p.167; Marquis and Allen, 1966, pp.1150-1160). The technicians who constitute nearly half of the population in the present study have hardly been investigated for their Information Seeking Behaviour (ISB).

As far as space technologists are concerned there are not many user-studies. Some studies (Allen, 1966, 1970 and 1977) about outside industries working on contract basis with NASA (National Aeronautics and Space Administration), an evaluation of GSFC library (Herner *et al.*, 1979), DOD (U.S. Department of Defense, 1965) user needs study and a study almost concurrently done with the present study by Raitt (1984) about scientists and engineers of ESA (European Space Agency) together with those from inter-governmental and international organisations in Europe are some of the related studies.

As mentioned earlier, it is important to note the contingency nature of findings of user-research. Importing findings of user-studies of an alien culture or country would not only be inapplicable, but might also mislead local professionals. It has been observed that most of the user-studies are done in the western culture with either American or British samples. Experts caution in applying the findings of such studies for other groups (Wilkin,1981, p.21; Meadows, 1974, p.108; Ford, 1973, p.91 and Ford, 1977, p.18).

Eventhough considerable research has been done in the area of user-studies in developed and information rich countries, especially U.S.A., Britain, Western and Eastern European countries, practically nothing substantial is done in less developed countries in general and India in particular. Not even a single major user-study in any of the areas of science and technology has been done in the country. Even the major national information system designs were not preceded by systematic, reliable and comprehensive user-studies. As such many present systems are surviving under 'symmetry of ignorance' (Kunz *et al.*, 1977, p.67-68), mathematical induction of convenient good or bad things, subjective considerations and idiosyncracies of people involved in planning information systems. Other systems are thriving under dubious reasons of avoiding the duplication of research and resultant economic benefit arising out of information systems but the hidden apparent reasons of an information system, as Myers (1970, p.26) says, could be treating information system as prestige centre or a form of fringe benefit to employees or a paid up insurance to serve an unanticipated need or to sublimate the feeling that one does not read or a centralised way of budgetary control over expenditure on documents. Saracevic (1980, p.226) discussing the information-need in less developed countries points out the absence of user-studies and says ". . . there is a dire need for conduct of scientifically based studies that will help define information needs in given developing countries."

Space science research in India began more than twenty years ago and space technology has more than ten years history in India. The Department of Space (DoS) is one of the youngest, up and coming departments of the Government of India. It consumes more than one-eighth of total R & D expenditure of the country. With its various establishments throughout the country and more than 14000 personnel, DoS has many formal information centres and libraries though not integrated formally into the nationl information system. The DoS has many ambitious plans for the years to come with many challenges for peaceful exploration of space. Hence the time is ripe to embark on scientifically planned user-studies to cover the young, energetic, ambitious and forward looking space technologists. As a beginning in this direction, this investigator, being the librarian of the youngest centre of Indian Space Research Organisation (ISRO), took up this base-line study of ISB of the space technologists working at the ISRO Satellite Centre (ISAC), Bangalore.

1.3 Objectives and Hypothesis of the Study

Objective 1 :
The Main objective of this study is to probe the information-transfer process and to determine the ISB of the Indian Space Technologists (IST). This simplified objective could be expanded as follows. To determine the motives and purposes of seeking information, nature and type of information required, degree of dependence on various sources of information, time spent in information-gathering, sources of bibliographic information used, whether information-gathering is delegated, if so, its nature, reasons for delegation as well as non-delegation, inter-personal information-sharing, informal communication-network and its nature, communication-behaviour, and the use of and interactions with the library by the IST.

Objective 2 :
The second objective of the study is to determine the user-characteristics of the IST. They include selected demographic, personality, organisational/ environmental and professional characteristics. These characteristics as variables of information-behaviour are intended to be used to stratify the space technologists as users of information.

Objective 3 :
The third objective of the study is to examine whether the various aspects of the ISB as described in objective 1 have any relation with the selected user-characteristics described in objective 2 and if such a relation does exist, to determine wherever possible the direction and magnitude of the correlation/ association.

The ultimate objective of the study is to assist in designing, operating and maintaining an information and communication system for the IST in which the inherent capabilities and limitations of various components of the system including sources of information best match with the requirements, attitudes and behaviour of the IST.

Hypothesis of the Study

The positive research hypothesis of the study (in relation to objective 3) is that ISB of the IST has a definite relation to the characteristics of the IST. The specific null hypothesis to be tested in the study is that there is no significant association/ correlation, other than due to chance, between ISB of the IST and their status, experience, nature of work, qualifications, specialisation and professional activities and achievements based 'performance'.

1.4 Scope and Limitations of the Study

1.4.1 *Scope of the Study*

An attempt is made here to present the scope of the study in terms of exclusions and elaboration of inclusions as well as resultant limitations of the study. It is very clear that "The purpose of information systems and their networks is to establish *communication* among people who share a concern about similar problems. Documents are only one means to this end" (Kunz *et al.,* 1977, p.65). Further "communication cannot be viewed merely as librarians work . . ." (Weinberg, 1967, p.41). As such the scope of this study is neither restricted to documentary and formal information-sources nor to the Library of ISRO Satellite Centre.

The scope of information is however restricted mainly to the current and future work-related scientific and technical information of the space technologists. But all such information which deal with professional competence and other organisation-related activities, though not specifically related to the present work, are also given due importance. Largely the social and personal information (as against work-related information) are excluded from the study.

The studies named variedly as 'user-studies', 'information-need studies', 'use-studies', 'information-transfer studies', 'communication-behaviour studies', 'Information-behaviour studies', 'dissemination and utilisation studies' are closely related, and often not precisely defined. Menzel (1966, p.43) tries to delineate these studies and says "when approached from the point of view of the scientist or technologist, these are studies of scientists' communication-behaviour. When approached from the point of view of any communication medium, they are use studies. When approached from the point of view of the science communication system, they are studies in the flow of information among scientists and technologists." Hence the terminology depends much on the approach and the angle from which one sees. To give a specific focus to the present study a simple linear approach to various information-related and information-seeking activities of users (as in Fig. 1.1) is assumed (on line with models of human communication and motivation theories). As per this assumption some motives and purposes of users give rise to information needs and requirements. To satisfy such needs and requirements users adopt many ways and means of accessing and searching sources of information. Then they try to acquire necessary information both regularly and in an adhoc way as necessary from these sources. In the act of acquiring information, the user uses or calls upon the sources of information and applies the information to the purposes as predetermined, thus leading to satisfaction or dissatisfaction in relation to purposes and needs. In reality these sequentially represented steps are neither discernible nor strictly linear. Though information-seeking behaviour in one sense deals with actual seeking of information represented at the stage of means of searching various sources of information, it would be more appropriate to cover other earlier stages and later stages to get a clear picture of the entire sequence. However, more emphasis is placed on the first four phases.

The actual application of collected information, either to solve a specific problem on hand or for furthering understanding or for knowledge in the area, is also equally important in understanding the information-seeking behaviour. However, on account of practical difficulties in carrying out studies of how information is applied to work, this aspect is kept outside the scope of the present study. Consequently, the satisfaction or dissatisfaction of users about information system could not be covered in the study. No attempt is made in this study to evaluate various sources of information as well as information used.

The communication-behaviour in this study is treated as a complementary activity to ISB and communication behaviour of the space technologists is included to a limited extent in case of formal communication-behaviour and to a greater extent in case of informal communication-behaviour in the informal communication network within the organisation.

Each user is an unique individual and differs from others in many ways. However, looking for generalizations and regularities among a set of users having common characteristics is necessary for a theory or discipline to grow. Kunz and others (1977, p.65-66) explain this problem

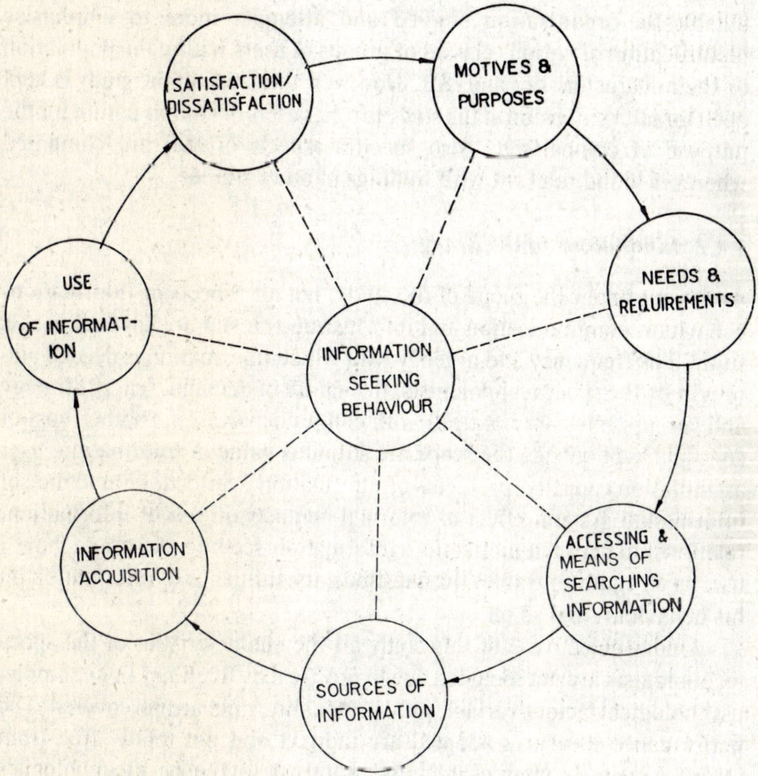

Fig. 1.1: Scope of Information Seeking Behaviour

of uniqueness versus the necessity of classes and say "On the one hand, information systems are always designed for a *class* of problems (and people). On the other hand, a problem (and a user) is the more unique, the more important it is". They further stress the importance of uniqueness and say ". . . the beloved ideal of universal compatability among information systems should be abandoned as the overriding principle . . ." As such, the present study aims less at generalising the results of information-seeking behaviour of the space technologists to outside-the-organisation studied and attempts more to emphasise identification of various classes or groups of users within the population by their characteristics and ISB. However, the scope of the study is kept open for later extension of the study to other users of similar nature for the purpose of comparison. Also specific aspects of ISB are compared wherever found relevant with findings of other studies.

1.4.2 *Limitations of the Study*

As set out under the scope of the study, not all aspects of information-behaviour, communication and information-transfer are covered in this study. The frequency and urgency with which information is needed and sought by the space technologists, the nature of demand, search strategy and success rates are not dealt with comprehensively. Another host of concepts kept outside the scope are stimulus value of information, user assimilation capacity, problems of information overload, redundancy of information system, effect of informal contacts on use of information, relationship between motivation, information-seeking and usage, how a user as an enquirer frames the questions, the influence of information on his behaviour, and so on.

Under objective 2 of this study all the characteristics of the space technologists are not included due to practical difficulties. For example, psychological factors/variables and personality traits are not covered. The performance measures adopted are indirect and not totally free from errors. Among the many characteristics and roles of the space technologists excluded are research guide's role, social roles, semi-official roles such as roles in Staff Association and other bodies. Among the factors which condition the information-behaviour of users who stand at the centre of ten systems proposed by Paisely (1968, p.3-6) social, political and economic systems are not considered in this study.

Relating to objective 3 of this study, the aim is not to find out cause and effect relationships among characteristics of the space technologists and their ISB, but to show association or correlation, if any. As Rowley and Turner (1978, p.58) point out "surveys can rarely establish a causal relationship, they can only indicate association or correlation." Further the inter-correlation among various aspects of information-behaviour, multiple and partial correlation of information-behaviour with two or more user-characteristics and factor and cluster analysis of information-behaviour as well as user-characteristics are not attempted in this study mainly due to limitations of resources. Even the inter-correlation among user-characteristics is done for selected characteristics and any correlation or association test in this study is limited to two variables/attributes at a time. One of the difficulties of carrying out above types of analysis is that the information-behaviour data is an admixture of nominal, ordinal and interval types (*i.e.* both qualitative and quantitative types).

Another limitation (normally associated with any user-research) is that it is not participative user-research with simultaneous system development but almost exclusively concerned with existing system and views about future information facilities based on prevailing information habits (Kunz *et al.*, 1977, p.68). Hence, to a large extent, respondents are conditioned by what is known to them and what is available to them in the present system.

The study is 'applied' and not 'basic' user-research and hence as far possible speculative answers to 'why a user behaves as he does' are avoided. The methodologies adopted in this study do have certain limitations. But specific aspects have been studied in depth using different methods and viewpoints. Coherence of all methods and aspects is maintained to meet the objective.

1.5 Organisation of the Study

This study was taken up in January 1982 and carried out till the close of 1985. The entire research, together with its outcome, is presented in this book in ten chapters. The present chapter deals with the need, objectives, hypotheses, scope and limitations of the study. As mentioned in the preface, review of relevant literature is presented elsewhere in greater detail (Sridhar, 1990).

The design of the study including methodology, data collection tools and data analysing techniques are presented in Chapter 2.

Chapter 3 presents an analysis of the characteristics together with general background of the IST.

Chapter 4 presents an analysis of motives and purposes of seeking information by the IST followed by correlation of motives and purposes with selected user-characteristics.

Chapter 5 presents the information requirements of the IST and correlation of nature and type of information required with selected user-characteristics.

How the IST have accessed information, the ways and means of their acquiring information are dealt in Chapter 6. The specific aspects presented are sources of bibliographic information, delegation of information-gathering work, time spent in information-gathering and accidental and direct acquisition of information and their correlation with selected characteristics of the IST.

Chapter 7 deals with the dependence of the IST on various information sources—both formal and informal sources. A relation of dependence on information sources with user-characteristics is also explored in this chapter.

Chapter 8 presents an analysis of communication networks and the communication-behaviour of the IST. In particular this chapter examines the extent of information-sharing among the IST, informal communication-behaviour and institutional and individual sources of information outside—the organisation, publishing, citing and co-authorship patterns and a trend of how these factors are related to selected user-characteristics.

Chapter 9 is concerned with how the IST use library documents and interact with the library together with the correlation/association of use of library documents and user-interactions with selected user-characteristics.

Chapter 10 presents a summary of findings, their implications and suggestions for further research based on analysis of the data and results of the study.

Towards the end are provided the bibliographical references as per Modern Language Association style manual (with slight modification) and appendixes wherein depicted additional data which are referred to in the body of the text.

1.6 Terms and Concepts Used

This being a social science oriented research many terms and concepts do not have precise definitions. Yet an attempt is made in this section to present the meaning, scope and operational definitions of certain important terms and concepts used in this study.

1.6.1 *Space Technologists*

The term 'Space Technologists' or IST in this work refers to all the technical staff other than Group-D employees, Medical, Transport and Library staff members of the ISRO Satellite Centre, Bangalore. In other words the inclusive definition of 'Space Technologists' refers to the scientists, engineers, managers, technocrats and technicians of ISRO Satellite Centre, Bangalore.[3] The technicians covered in this study have vocational training or diploma in engineering and technology trades.

1.6.2 *'User' and 'Use'*

The term 'user' in this study refers to the defined population of 'space technologists'. Hence it includes actual as well as potential users and non-users of ISRO Satellite Centre Information System.

The concept of use is often defined as the extraction of content from a message to meet a need. The operational definition of 'Use' for collecting data about the use of the library documents in this study is as follows. For inhouse use of documents Kent's (Kent *et al.*, 1979, p.61). definition of use as physical selection and the act of leafing through pages of document is adopted. For circulation or loaned use, each record of having lentout or renewed is considered as one externally circulated use of a document.

1.6.3 *'Information' and 'Communication'*

Both the terms have already been discussed at length in the previous sections. However, the context of their use in this study is presented here. The term 'Information' is used as a noun in this study, meaning contents of a message, irrespective of its manifestations and length of exposition. A widely used definition of information is that it is a stimulus that causes change in the degree of certainty or uncertainty of the receiver. For the purpose of this study 'information' is that which adds to knowledge and/or helps one in decision-making and problem-solving situations. Hence it covers the knowledge, facts and data relating to work and the profession/discipline of the subjects and programmes and activities of the organisation but excludes routine administrative, social and personal information. In terms of the manifestations 'information' includes oral and written, documentary and non-documentary, statistical, pictorial, graphic, descriptive, bibliographic, and other forms from formal and informal sources.

'Communication' essentially refers to a process and a simple and generic model of communication involves a sender, message, coding, channel, decoding and a receiver. Such a single channeled, simple, linear and unidirectional communication model between two individuals is an over simplified model. But it serves the purpose of this study. Communication is more than mere contact. It is not even mere information-transfer. Being a higher-level phenomenon, communication involves both information—transfer as well as a change in the state of the receiver and sender of message.

1.6.4 *Information-Seeking Behaviour*

The term 'information-behaviour' is used here to include all activities comprising information-seeking, information-gathering, information-receiving and communicating. Information-seeking precedes gathering and information-gathering most of the times implies seeking. At the outset information-gathering appears more as a physical act than seeking. However, in a broader interpretation information-gathering includes and presupposes information-seeking and hence is a broader concept than information-seeking behaviour. However, both the terms are quite often used interchangeably in the literature and 'information-behaviour' includes both.[4] A very practical and broad definition of ISB is given by Auster (1982, p.178) as "the field composed of studies that are concerned with who needs what kind of information and for what reason; how information is found, evaluated and used; and how these needs can be identified and satisfied". A supporting analysis of the meaning and scope of this term is given by Wilson (1981, p.3-15). ISB results from the recognition of some need as perceived by the user. He calls that the study relating to search strategies in the following information-seeking paths as the study of ISB :

(i) search paths by a user independent of any information system,
(ii) search paths involving either a mediator or an information system's technology,
(iii) paths employed by a mediator to satisfy a user's demand for information and
(iv) paths employed by a sophisticated technology on behalf of either the user or the mediator.

The motives for investigating search processes may be to make inferences about need, or it may be to uncover facts relating to other variables related to design, development or adoptation of information systems (Wilson, 1981). Wilson further distinguishes both basic research and applied research aspects of ISB and says "... the study of information-seeking behaviour can stand on its own as an area of applied research where the motive for the investigation is pragmatically related to system design and development... [and] why the information seeker behaves as he does . . . is an area of basic research. . ." (p.7) in which the present study is not concerned very much. Wilson further says ". . . before a generally applicable theory of ISB can be evolved, the context of the research must be narrowed so that crucial determining factors can be identified and analysed" (p.11). Krikelas (1983, p.6) gives a very simple definition of ISB ". . . as any activity of an individual that is undertaken to identify a message that satisfies a perceived need." As such the meaning and scope of ISB in this study is as follows : The motives and purposes of seeking information, the nature and type of information required, the ways and means of accessing, searching , identifying and acquiring work-related information, the degree of dependence on sources of information, communication-behaviour, use of library and user-interactions with the library.

1.6.5 *Need, Want, Demand, Use and Requirement*

'Need' is a more abstract and difficult-to-define concept. Both Ford (1980) and Krikelas (1983, p.6) define 'information need' as an awareness or recognition of not knowing or existence of uncertainty. The draft definitions of 'need', 'want', 'demand', 'use' and 'requirement' in relation to information proposed by Brittain (1971) and supported by Line (1974, p.87) are quite adequate for the purpose. *'Need'* is what an individual *ought* to have, for his work, his research, his education, his recreation, etc. A need may or may not be identified as want.

'Want' is what an individual *would like* to have, whether or not the want is actually translated into a demand on the library. Individuals may need an item they do not want, or want an item they do not need. A want, like a need, is a potential demand.

'Demand' is what an individual asks for; more precisely, a request for an item of information which is believed to be wanted. Individuals may demand information they do not need, and certainly need or want information they do not demand. Demand is partly dependent on expectation, which in turn depends partly on existing provision of library or information-service. A demand is a potential use.

'Use'[5] is what an individual actually *uses.* A use may be a satisfied demand, or it may be the result of browsing or a chance. Individuals can only use what is available. Use is, therefore, heavily dependent on provision and availability of library and information-service. A 'use'

usually represents a need of some kind. But 'need' is independent of 'use'. Uses can be partial indicators of demands, demands of wants, and wants of need. Identification becomes progressively more difficult from the 'hard' use to the often nebulous and unstated need.

'Requirement' is a useful bridging term; it can mean what is needed, what is wanted, or what is demanded, and can therefore be usefully employed to cover all three categories. But the term 'requirement' is closer to the term 'need'. Many studies of 'needs' have infact been studies of requirements. It is in this sense that the term 'information-requirement' of the space technologists is employed in this study.

Wilson (1981, p.8) presumes that ". . . as part of the search for the satisfaction of needs, an individual may engage in ISB." The interesting and most practical aspect is that ". . . needs arise out of the roles an individual fills in social life. So far as specialised information system is concerned, the most relevant of these roles is 'work role', that is, the set of activities, responsibilities, etc., of an individual, usually in some organisational setting in pursuit of earnings and other satisfactions" (Brittain, 1971, p.9).

1.6.6 *Formal Sources and Informal Sources*

Another set of terms which are often used with lesser consistency are formal, non-formal, semi-formal, informal, oral, documentary, personal and inter-personal sources of information. The terms extensively used in this study are 'formal and documentary sources' of information and 'informal and inter-personal sources' of information. While the concept of 'formal and documentary sources' is quite clear, the 'informal and inter-personal sources' is defined to include oral, personal (inter as well as intra) and selected semi-formal and nonformal sources depending on their degree of inter-personal and informal nature.

NOTES

1. Compared to information, knowledge is believed to be passive and an item of knowledge becomes an item of information only when it undergoes the active process of communication.
2. Brittain (1975, p.429) says a lip service has been paid to this objective in many user-studies.
3. The term 'space technologists' is preferred to 'satellite technologists' in this book due to its popularity and the commonalities of 'satellite technology' with other component areas of 'space technology'.

4. Krikelas (1983, p.8) feels ISB is concerned with activities associated with satisfying immediate needs and information-gathering behaviour is concerned with activities associated with deferred need.
5. An operational definition of 'use' has already been presented earlier and it is in this context that the present study emphasises work-related ISB. This study is concerned with the present as well as future 'work-related information-needs' of the Indian Space Technologists.

2

RESEARCH DESIGN AND METHODS

2.1 Design of the Study

An exploratory-survey approach is adopted in this study with local institutional orientation like a case study. A cross-sectional design (as against retrospective design) with emphasis on norms of critical incident technique and occasionally on longitudinal approach and activity sampling technique is followed. The structure of communication-network is looked at from a structuralist approach (Adam, 1982, pp.396-397). The design of the study began right in the beginning of 1982 with design and testing of data collection instruments followed by collection of data mainly during 1983 and 1984.

2.2 Methodology and Data Collection Tools

Several complementary methods of investigation as well as data collection tools are used with the intention of gaining greater insight and more clear and complete picture of the space technologists and their ISB than a single method (Blagden, 1980, p.35). Multiple investigation methods (both direct and indirect methods) and multiple data collection tools are used in a helpful sequence so as to enable cross checking of data for its validity, reliability and consistency.[1] Sequencing of methods is such that each method tries to gain from and supplement the previous methods. All methods, techniques and approaches are directed toward the core problem and have enabled gathering data about the same set of users by different techniques depicting different perspectives such as users' opinions, researcher's observation, results of discussions, self reported activities of users and analysis of historical records.

The questionnaire method used extensively is supplemented by semi-structured personal interviews, a self-reported weekly summary of information-related activities (a modified diary-keeping method), direct unobtruisive as well as indirect observation, analysis of records, quasi-experimental method and a host of indirect methods including observation, recording and examination of records of various user-interactions with the primary library, use of library documents and a study of communication behaviour through publishing, citing and co-authorship patterns of the space technologists.

Main dependence on questionnaire method is to have maximum cost efficient coverage for quantitative research with available resources and due to practical difficulties of user-resistance and getting time and appointment ofbusy space technologists for interviews and diary methods.[2] The costly observation method is employed for limited purposes such as how space technologists visted and used primary library, consulted card catalogues, etc. The qualitative data has been collected by summary sheet, interview and questionnaire (open ended question) methods.

Data collection tools were developed simultaneously but tested and administered in a sequence so that they can be made to supplement each other and design of one tool can be improved from the result of the tool administered earlier.

2.2.1 *The Weekly Summary Sheet*

Having found no systematic personal diary keeping habits with the respondents, a slightly modified 'summary sheet' providing information-related activities of respondents over a week has been designed after a predesign and exploratory discussion with selected users and after a pilot test with three respondents (Appendix-I). Summary sheets were administered to a 20 per cent systematic random sample (of the population arranged in the ascending order of staff numbers as of December 1982) for a six-week period from 2 January 1983 to 12 February 1983. Inspite of personal as well as telephonic reminders, persuasion and clarifications of doubts, the response was abnormally low. Out of 150 persons contacted only 20 (13.3%) have fully responded and another 22 (14.7%) have given partial responses *i.e.*, summary sheets of 1 to 5 weeks were returned. In all, out of 900 man-weeks of summary sheets administered 177 (19.7%) man-weeks of summary sheets were returned duly filled. The response sample was not very much representative of the population and was not adequate for a longitudinal study. Responses appeared to be biased towards information-oriented and intensive users of information. As the respondents did not resist disclosing identity the data received through 'summary sheets' has not only helped as a starting point for further data collection though questionnaire and interviews but also to relate it to the data from other methods.. A probe into the reasons for low response to 'summary sheets' indicated that lack of time, especially for those at higher level, lack of motivation (some were sceptical of the study), laziness, hesitance for discussing personal activities for fear of exposure of bleak information-related activities and lack of need for technical information (as compared to MIS type of information in case of some of the space technologists) as major reasons.

2.2.2 *The Questionnaire*

Keeping in view the scope and objectives of the study, environment of the IST, results of observations, response pattern to 'summary sheet' and questionnaires of similar earlier studies, a comprehensive almost structured questionnaire (Appendix-II) was designed based on prepilot investigations and results of a pilot study with 20 respondents. It was a largely self-administered questionnaire seeking facts about user-characteristics (objective-2) as well as user-opinion, attitude and preferences relating to information-behaviour (objective-1). All questions, except one or two, are partially closed-end questions with provision for cross checking responses within the questionnaire as well as with data from records. Yet repetitions have been kept to a minimum. The questionnaire which needed about 40 minutes to complete brought mostly quantitative data which was not available from other sources/methods. It carried a five-point scale-rating for selected questions and assumed that the points on the scale are at an equal distance for the purpose of analysis. The rating scale was conceived in the ascending order with nomenclature for each point on the scale and hence the numerical values for judgements of respondents were not expected to differ from one another. Further, the validity and reliability of the questionnaire has been checked. In order to ease the quantitative analysis it was found necessary to have a scale of five points and responses on scale is better than binary responses.

The qestionnaire was administered to all the 807 persons in the population (except 13 who have refused or were not accessible) with the hope of getting adequate and representative sample for analysis of all aspects of ISB and user-characteristics as planned. A systematic follow up over 10 months (July 1983—March 1984) and personally filling the questionnaire in some cases as an adjunct to interview and discussion resulted in 68.5 per cent response including 9 (1.7%) rejections/invalid responses and 30 (6.3%)

anonymous responses. A followup of non-respondents did not reveal any significant pattern. There was more than 90 per cent consistency between the two responses provided by 10 selected users after a gap of six months.

2.2.3 *The Interview Schedule*

The interview method was mainly to followup and have insight into certain qualitative aspects and issues not covered in the questionnaire method. The interview or discussion with selected space technologists was done at three levels. The questionnaire itself was used as a fully structured interview schedule and filled by the investigator in as many as 50 cases. Secondly, many responses to the questionnaire and certain post-analysis issues were picked up for further discussion and/or clarifications with over 50 cases. Thirdly, an independent semi-structured interview schedule (Appendix-III) designed and developed along with questionnaire and tested in a pilot study was used with a purposive sample of 30 space technologists for an hour-long discussion with each. The sample included few non-respondents to the questionnaire. Information gathered by this method is used as a supportive evidence in the analysis of the data collected through questionnaire.

2.2.4 *The Observation Method*

A semi-structured observation method is followed in this study about how the space technologists have visited the library, moved in various parts of the library, consulted card catalogues and used current journals inside the library. Apart from these direct observations, some indirect observations have been made through records of interactions with the library and certain outputs of the space technologists like publications.

2.2.5 *Analysis of Records*

First, the personnel information file is used to enlist the population with user-characteristics. Secondly, library records like records of user-suggestions/requests for documents, procurement intimations sent by library to users, reservations made on documents by users, reference queries answered by the library, and reprographic indents made on library by users during 1982 have been extracted and analysed. In addition a three-month circulation data has also been elicited for cross checking on results of quasi-experimental method of document-use study.

2.2.6 *Quasi-Experimental Methods*

Controlled experiments in the true sense were not possible except in the study of document-reservation behaviour and as such quasi-experimental study of circulation use of a 20 per cent collection sample of ISAC Library books, reports, standards and current journals over 10 months and a three-month circulation sample of two major outside libraries, namely IISc (Indian Institute of Science) and NAL (National Aeronautical Laboratory) Libraries were carried out. Most of the studies under the quasi-experimental research methods are carried out as independent case studies.

2.2.7 *Indirect Methods*

In addition to above methods, many independent indirect methods like studying publications of the space technologists for their publishing, citing, co-authorship and collaborative research patterns are also employed.

2.2.8 *Other Methods*

Lastly, it is pertinent to observe here that many other research techniques or methods used in this study are only in the form of normative principles rather than as independent methods. Some such important methods are the critical-incident technique, longitudinal approach and activity sampling technique in data collection.

2.3 Validity and Reliability of Data Collection Tools

Since a large amount of quantitative data has been elicited through the questionnaire constructed for the purpose, it was necessary to examine how far this instrument was valid and reliable (Ley, 1972, p.128-129; Webb and Campbell, 1973).

While a predesign test and a pilot test of the questionnaire have taken care of the face validity of the tool, the content validity will be discussed as adequacy and representativeness of sample in a subsequent section and in chapter 3. The empirical validity has been worked out through out this thesis in terms of validity of associations or correlations. A construct validity is possible only in a study where a network of statements predicts relations among observables.

The reliability of the data collection tool essentially depends on how far consistent responses could be elicited, given the same tool to the same individual. Two ways of checking reliability is comparing data collected by two different methods (to some extent this method is also adopted in this study to invalidate inconsistent data and responses) and comparing data elicited through the same tool at two different times.

A simple test about the reliability of the questionnaire was conducted by administering the questionnaire for a second time to ten carefully selected representative samples after six months from the date of the earlier response. Six months was judged to be an optimum time by which the variations in response due to change in the attitude/behaviour of the respondents as time passed and the chances of the respondents memorising what they have answered on earlier occasion were minimum. An examination of the responses revealed that a minimum of 7 out of 10 respondents have maintained consistency in their replies to any sub-question and it varied upto 10 out of 10. The four important questions namely, question numbers 1,2,3 and 11 where data was sought on a five-point scale, the percentage of consistency estimated (by ignoring the sign of error due to inconsistencies and taking the mean difference for a five-point scale) was 93 per cent, 92 per cent, 97 per cent and 92 per cent respectively. If the sign of error due to inconsistent responses is considered, they cancel out each other to some extent and the overall consistency would still improve.

2.4 Data Analysing Techniques

The wealth of data (quantitative) amassed by adopting various methods of investigation and data collection tools have been checked, coded, transferred, validated for consistency, counter checked and tested for reliability before processing and analysis. The data handling, validation, processing and analysis have been carried out using a computer. Yet checking, tabulating and calculation of statistical significance tests for association and correlation have been partly carried out manually. Finally a machine readable data base has been developed for processing and analysis. In the process, nine responses have been invalidated and rejected for obvious and glaring inconsistencies.

2.4.1 *Checking and Validation of the Data*

The entire data about each user in the population drawn from various sources and tools such as population list, questionnaires returned, data about interactions of users with information system initially kept on a 5"x 8" card were transcribed by the investigator personally, with the hope to minimise errors, on to coding sheets. Upto six Hollerith cards have been punched per user from these data sheets. All the cards about a user were logically tagged together with the secret user code at the beginning of each card followed by the card designation.

Some user-characteristics which are available from records and other sources including population list were individually compared with responses to questionnaires based on staff number or secret user code unless user has opted to be anonymous. Even in case of nonresponses to questionnaires, the available data was transferred to machine readable form.

Transcribing and keypunching errors of the data and inconsistencies in response to questionnaire have been checked through data validation programs and extensive manual error checking.

2.4.2 *Analysis of the Data*

Since this study is concerned with a fairly exhaustive population of a single organisation distribution of data among entire population as well as response population is presented wherever possible and necessary.

Leaving the descriptive, illustrative and qualitative data obtained mainly through interview, open ended question in the questionnaire and weekly summary of information-related activities, the rich and wide variety of quantitative data about each individual in the population is analysed statistically.

The number of aspects of ISB and the number of user-characteristics were too many to be individually considered for correlation/association with each other. At the first level all aspects of ISB and user- characteristics have been broadly grouped. Existence of correlation/association among user-characteristics themselves have been statistically checked. Based on such correlations/ associations and the importance of user- characteristics, six important characteristics highly associated/correlated with other characteristics of users have been chosen for further correlation /association with various aspects of ISB.

2.4.3 *Statistical and Mathematical Techniques Used*

The data gathered were both 'qualitative' and 'quantitative'[3].

Apart from simple and common statistical techniques such as presentation of frequency and contingency tables, percentages, pie and bar charts, some statistical tools like mean, standard deviation, measure of association, measure of correlation, etc. have been used both for the data as well as for the composite indexes such as library-use index, library-interactions index and professional activities and achievements-index. The construction of indexes are explained in more detail at appropriate places.

The scaled values in the questionnaire responses have been converted to composite scores and the resultant weighted mean is presented as an index together with standard deviation. The formulae used are :

$$\text{Mean,} \quad \bar{x} = \frac{\Sigma x_i f_i}{\Sigma f_i}$$

$$\text{S.D. (Standard Deviation),} = \frac{\Sigma \ (f_i x_i^2)}{\Sigma \ f_c} - \bar{x}^2$$

where, x_i = Value from the scale

f_i=Frequency of occurrence of a value

The objective 3 of this study is to findout correlation/association, if any, of selected user- characteristics with ISB. Whenever an association/ correlation (as the case may be) between any two user- characteristics or between a user-characteristic and an aspect of information-seeking behaviour has to be determined one of the following three statistical techniques has been employed depending on the type of characteristics involved.

(i) If at least one of the two characteristics is qualitative then chi-square (x^2) statistic is calculated based on the formula.

$$\bar{x}^2 = \Sigma \frac{(f_o - f_c)^2}{f_c}$$

where, f_o = Observed frequency

f_c = Expected frequency

The workedout chi-square statistic is checked against the tabulated value for a given degrees of freedom *(df)* and 95 per cent probability or 0.05 significance level *(p)* to accept or reject the association.

(ii) If the characteristics involved are either qualitative or quantitative but amenable for ranking, Spearman Rank Order Correlation Coefficient (r_s) using the following formula is worked out.

$$r_s = 1 - \frac{6 \Sigma D_i^2}{n(n^2 - 1)}$$

where, D_i = Difference between the ranked value of x_i and the ranked value of y_j

n = No. of ranks considered for both x_i and y_j

A t-test with the following formula is made on the result obtained for finding the statistical significance.

$$t = r_s \sqrt{\frac{n-2}{1-r_s^2}}$$

The value so obtained is compared with the t-test value from standard table for given degrees of freedom at 95 per cent probability or 0.05 significance level to accept or reject the correlation.

(iii) When both the characteristics involved are quantitative, Pearson's Product Moment Correlation Coefficient *(r)* is calculated using the following formula.

$$r = \frac{\Sigma \Sigma f_{ij} \ (x_c - x)(y_j - y)}{\Sigma f_j \ (x_c - \bar{x}^{-2}) \sqrt{\Sigma f_j (y - y)}}$$

where, x_i, y_j = Weighted or grouped values of characteristics involved

f_i, f_j, f_{ij} = Frequency of occurrence (of grouped values) of characteristics

$\bar{x}, \bar{y}$ = Mean values of the characteristics

A t-test is conducted on the resultant r as explained under (ii).

The resultant correlation coefficient values are interpreted in this study with the following assumptions: <0.2 as slight and almost negligible correlation; 0.2 -0.4 as low correlation and definite but small relationship; 0.4–0.7 as moderate correlation and substantial relationship; 0.7–0.9 as high correlation and marked relationship; >0.9 as very high correlation and very dependable relationship (Guilford, 1956, p.145).

Another important rather mathematical technique used in the sociometric analysis of informal communication networks is 'equivalence relation' (Sridhar, 1980). The method is very similar to 'snow ball' technique of social research. The network of informal contacts in the response population and of ISAC authors have been divided into mutually exclusive or disjoint clusters called equivalence classes.

2.5 The Population, Sample and Errors

2.5.1 *The Population*

Based on personnel information file of the organisation an exhaustive list of population was prepared and updated till mid 1983 with a cutoff point in June 1983, when actual questionnaries were distributed. Data from personnel information file and records of the library were coded, checked and added against each member in the population list. Members were assigned secret user codes.

2.5.2 *The Sample*

This is almost a census study of defined population of users. Data from observations and records have been collected about the entire population. In other words, most of the user-characteristics and all user-interactions with library are available for the entire population. The data from questionnaire is available for response population (68.5%). The non-sampling error due to non-response to questionnaire is taken into account in the analysis. The summary sheets have been distributed to a 20 per cent systematic random sample.

Though observations and examination of records for user-interactions with library and information system have been made about the entire user-population, the period and quantum of observations and data from records have been based on samples. Stratified purposive sample has been selected for data collection through interview. Apart from covering users of different strata, certain users who indicated some scope for further exploration through their summary sheet and/or questionnaire were selected.

Sampling techniques and sample frames (adopted for various

observations) and indirect methods are explained in the appropriate sections. As far as use of library documents are concerned a 20 per cent stratified systematic sample from the collection (called 'collection sample') for books and reports was adopted to record the number of times they were issued out to users in the population over a period of 10 months. In case of journals, two latest issues of current journals were monitored for three months from the date of their first display to record both number of times borrowed and number of times used within the library (to a maximum of four recordings in a working day in case of inhouse use). Further, a three-month sample of circulation records (called 'circulation sample') were scanned to record how each individual in the population has borrowed books, reports and journals from the library.

2.5.3 *Errors and Biases*

The sampling error of the study is almost negligible as full population is covered by the main data collection methods namely, questionnaire, analysis of records, and indirect observation through quasi-experimental methods. In other methods of data collection especially indirect methods, adequate and representative sample have been employed. A biased purposeful small sample for interview and a systematic random sample for summary sheet method with low response do not contribute for sampling error as they were mostly qualitative data of illustrative cases.

The extent of non-sampling errors due to non- response is obviously negligible as we could see in Chapter 3 (Characteristics of the Indian Space Technologists) that non-responses do not have a pattern. Adequate and representative response has been received for questionnaire from all categories/groups of users. Further, for the questionnaire method there has been a reasonable response rate of 68.5 per cent. As a first step against other non-sampling errors built in checks in the questionnaire and extensive checking against data from records and observation methods have been employed. All doubtful responses have been, telephonically, clarified by the respondents.

Lastly, respondents were given option about disclosing their identity (i.*e.*, staff number) in the questionnaire so as to relate data from questionnaire with the data about their characteristics and interactions with the library. To check on any possible bias in case of those who disclosed identity, an analysis of 30 anonymous responses are separately made and compared with the total response throughout the analysis. The anonymous responses do not show any significant variation from the total response indicating, thereby, absence of bias due to identity disclosure.

NOTES

1. 'A study which has recourse to several sources is reckoned to portray a more rounded image, although the degree of compatability between [them] . . . is still not known' (Rowley and Turner, 1978, p.62.).
2. The weekly summary sheet and interview methods have resulted in low response.
3. The terms 'qualitative', 'quantitative' and 'characteristics', in this section are used in the statistical sense.

3

CHARACTERISTICS OF THE INDIAN SPACE TECHNOLOGISTS (USER-CHARACTERISTICS)

3.1 Introduction

No user-community of an information system is fully homogeneous. Though a majority of the users of a system would come together for a particular purpose and are comparable by one or two criteria, they are divided among themselves by many individual characteristics. As noted earlier, the knowledge of the population being served by an information system is an essential requirement for providing useful services. Just like understanding the user is half the battle in providing information-services, knowing the structure and composition of the user-community in terms of various characteristics by which they can be compared and contrasted is half the task in understanding the users.

The purpose of the present chapter is to describe the characteristic features of the population of this study, namely, the Indian space technologists. The first part highlights space activities in India, particularly the structure, functions and other background of the ISRO Satellite Centre (ISAC), the general nature of work of the space technologists and finally the 'primary library' of the population, namely, the ISAC library in terms of its growth over years and its present status. The quantitative aspects of user characteristics collected from the records of the organisation including that of ISAC library and through the questionnaire are presented in the second part. Whenever data are known from more than one source, a comparison is presented to check the reliability of the data. When a particular data is collected from records as well as from questionnaire a comparison is made of how a particular characteristic is distributed among the entire population as well as the response population to emphasise the representativeness of the response population.

The characteristics of users are innumerable and there are several ways of grouping them. Vickery (1973, p. 45) apart from identifying many characteristics of users as recipients of information has emphasised that institutional environment and work activity (job) have major impact on ISB. This viewpoint has been further reiterated by Slater and Fisher (1969, p.v) in their study of use of technical libraries. The three clusters of factors which affect user's utilisation of information are psychological factors, effectiveness of available services and characteristics of the user and his environment.

Of the many characteristics considered in the earlier user-studies, those which are deemed relevant, important and appropriate for the present study are discussed here. User-characteristics have been grouped into several ways like internal and external to the individual, sociological, psychological, organisational, etc. The grouping of user-characteristics in the present study is not mutually exclusive but oriented to the purpose and convenience. Further, many of the user-characteristics have a very high association or correlation as discussed later in this chapter irrespective of the group to which they belong. It is neither feasible nor desirable to correlate every characteristic of users with every aspect of ISB. Hence, after a preliminary examination and inter-correlation of user-characteristics among themselves, certain characteristics have been chosen to correlate with selected aspects of ISB based on observation of the data and inter-correlation/association of user-characteristics.

3.2 The Indian Space Technologists : General Background

3.2.1 *Space Activities in India*

India is the seventh member of the World Space Club and the First Third World country to have entered the operational era of space technology with independent and advanced space programmes. It has the third largest scientific and technical manpower.[1] It is estimated that "out of 64,875 personnel directly engaged in R & D in (India), 42 per cent were engineers, 38 per cent were natural scientists, 16 per cent were agricultural scientists and 2 per cent each were medical and social scientists" (India, DST, 1982, p. 5).

By and large, the expenditure on science and technology in India has increased rapidly since independence and in the sixth plan it amounted to about 2 per cent of all public expenditure by the States as well as the Central Government (Nature, 1984, p. 584). The R & D expenditure itself is estimated to be 0.66 per cent of the country's GNP and 86 per cent of which is borne by the government — both Central and States and the rest by private sector. Within government's contribution, about 90 per cent is from the Central Government and 10 per cent from the State Governments (India, DST, 1982, p. 3). The DOS, one of the youngest S&T oriented departments account for 13.3 per cent of total R&D expenditure of the major scientific agencies (p. 18).[2]

The space research activities started in India in 1963 and the development and application of space technology gained impetus with the formation of the Space Commission and the DOS in 1972. The space commission is an advisory and policy-making apex body with responsibilities for framing the policy, formulating budget for DOS and implementing government's policy in matters concerning outerspace. The DOS as an executing body has the responsibility for all matters relating to space technology, space applications and space sciences. The ISRO is the primary agency of DOS charged with the responsibility of executing the R&D programmes and schemes of the DOS in accordance with the directives and policies laid down by the space commission and the DOS.[3]

3.2.2 *The ISRO Satellite Centre*

For the purpose of the present study, users have been exclusively, and fairly exhaustively, drawn from the ISRO Satellite Centre (ISAC). This enabled the investigator to examine in greater detail the background of the users in terms of activities, scope and nature of the organisation. ISAC is one of the four major centres of ISRO located in Bangalore with the responsibility mainly of (1) planning, detailed definition, design, fabrication, testing and integration of all ISRO space crafts, (2) development of technology related to satellite mainframe, (3) long-range planning of future Indian satellites. Thus, ISAC is the prime focus of ISRO's satellite programme.

ISAC had its beginnings in the then Indian Soviet Satellite Project (ISSP) (on 1st September 1972) at Bangalore with a small band of specialists.

It was also called Indian Scientific Satellite Project (ISSP) and ISRO Satellite Systems Project (ISSP). The first Indian satellite 'Aryabhata' built in this project was launched on 19th April 1975 at Soviet Cosmodrome.

Subsequently the project became one of the four full-fledged major centres of ISRO with the present name of ISAC on 3rd November 1976. Since its inception, the centre functioned from hired/temporary premises at Peenya Industrial Estate, Bangalore till it moved to its own premises at Kodihally near Bangalore Airport during May-June 1984.

DOS had the sixth plan outlay of the order of Rs.640 crores and of which Rs. 86 crores (*i.e.,* 13.4%) was the share of ISAC.[4] The fund distribution within DOS is approximately in the ratio of 7:2:1 among space technology, space application and space sciences.

DOS has a manpower of 15,133 as on 1st April 1985, of which about 66 per cent are technical staff and the rest, the administrative staff. The ISAC's share of manpower is about 12 per cent *i.e.,* 1834 as on the same date. Nearly 71 per cent of the staff of ISAC are technical (as against 66% in entire DOS), of which 92 per cent are other than Group-D employees.[5]

The work at ISAC is divided among various projects and divisions.[6] The projects are organised on inter-division/section and inter-centre basis with identified teams participating in the development tasks at subsystem/ system levels of spacecrafts. While the projects are responsible for the execution of specific time-bound tasks, the divisions are entrusted with the responsibility of R&D support required for current and future projects.

In addition to time bound projects, several Technology Development Programmes (TDPs) are undertaken each year by small groups within divisions, each one mainly centered around one or two individuals. The TDPs are low budget and slightly time relaxed R & D projects going side by side with main projects and they act as feeder systems to main satellite projects starting with new concepts and indigenisation of foreign technology.

In addition, technical personnel of ISAC are involved in various supporting technical activities, facilities and services such as environmental test facilities, mechanical fabrication facilities, electronics fabrication facilities, quality assurance, etc.

3.2.3 *Nature of Space Technology and Space Technologists*

It may be desirable, at this stage, to clarify the use of the terms 'space sciences', 'space applications' and 'space technology'. The space activities have been quite often referred as space science and technology to mean involvement of both science and technology. Space science and space technology have been, at times, used interchangeably. However, space technology has been the result of application of post Second World War research initially ". . . . fuelled by a combination of scientific and military objectives with national prestige playing an important role" (Dhawan, 1983, p. 4). "Over the years there has been a transformation from scientific exploration . . . . and national prestige and defence . . . . to the fulfilment of economic and other needs on earth (Dhawan, 1983, p. 8). Hence, it is aptly said that "space science and technology is neither a new science nor technology. It is, in fact, the application of space techniques to the already known science and technology on the earth" (Rao, 1976, p. 18).

The term 'space applications' in this context mainly refers to activities and work which reap the fruits of space technology by applying it to various areas of applications such as scientific, communication and remote sensing.

Sketching the nature of space technology, Dhawan (1983, p. 4-5) says —

> "space flight is primarily a matter of technology rather than science. It is true that the design and development of rockets and spacecraft draw heavily upon the knowledge of many sciences. But this complex combining of technical knowledge from many disciplines to achieve a set task in a very inhospitable environment is essentially an engineering systems job with management techniques which have evolved to deal with the large scale, complex and high risk ventures in which a large number of disciplines, individuals and institutions are interlinked."

(i) *High-Technology*: The space technology-activities are normally considered to be high-technology activities characterised by very large developmental efforts in state-of-the-art technology, multidisciplinary work, substitution or augmentation of mind power than muscle power, involving complexity and demanding high degree of reliability.

(ii) *Complex and Large Scale Activities*: The mission of space technology is not only complex, but is also a large scale one.

(iii) *R&D or Industry?* : Though the work is often described as R&D, it is less of research and more of development in nature. Hence, it may be more appropriate to classify it somewhere in between R&D and industry. Naturally, the organisational pressures and role strains are more in development than research-oriented organisation (Evan, 1962, p. 346).

(iv) *System Analysis and System Engineering*: By looking at the nature of the work in another way, it can be said that the work is predominantly system analysis and system engineering in nature, involving project management and reliability and quality control due to inherent complexities of the mission.

(v) *Heterogeneous Nature*: The teams are of a heterogeneous nature, in the sense that many specialists of different scientific and technical background work in the team. The subjects of specialisation of individuals are quite diverse. Yet all of them work toward a common mission/goal which needs a very high degree of coordination. Discussing the duality of the discipline and mission oriented research, Weinberg (1967,p. 42) says that NASA with the exploration of space as the mission covers almost all the traditional disciplines like chemistry, physics, astronomy, biology, etc. The duality is that the information-generated in the discipline is useful in the mission and *vice versa;*

(vi) *High Risk*: Since the responsibility of the centre is to make workable or operational satellites it involves a high risk unlike traditional R&D. In fact, R&D itself, by its very nature deals with an unsure future. But in traditional R&D, the cost of failure is much less than that in space technology. The high risk leads to a need for high reliability and stringent quality control as well as considerable redundancy and hence increased cost. Thus, there exists a trade off of cost, time and risk.

(vii) *Imported Technology*: Another corollary of high risk is use of imported components, parts, equipment and hence to some extent the technology itself.

(viii) *Time-Bound Nature*: Yet another feature of the space technologists is their involvement in time-bound projects. About 3-4 main projects are always on hand in the centre and by the time a project is completed the next would have already undergone preliminary stages of conceiving, studying, defining and proposing for approval. In addition, several TDPs will always be in progress at a single time interval. One important aspect of these two types of activities is that the scientists, engineers and other technical staff are changed periodically from a project to TDP on completion of a project and *vice versa*. It is also possible that one can be simultaneously involved in both types of activities.

(ix) *Staff Mobility and Turnover*: There exists a considerable staff mobility and turnover, especially in the case of juniors and less experienced staff. This is in addition to mobility due to redeployment of manpower within ISAC as well as within ISRO. Further, the staff strength of the centre has been on the increase during the study period. Table 3.1 depicting the growth of ISAC Library membership, indicates steep increase in staff strength of ISAC. An analysis of membership turnover in ISAC Library during the year July 1982 and June 1983 indicated that there was a 16 per cent increase in membership and about 4 per cent have left ISAC during the year. The total personnel employed at ISAC has doubled in almost five-years period ending 1983 and it is likely to be doubled once again for the next five years.

(*x*) *Relations with Industry*: Space missions need all sorts of hardware from a variety of industries in the country which are directly or indirectly involved in the mission. Space technology has not only acted as one of the catalysts for industrial growth but has also given many spinoffs to industry. About 100 industries have worked with ISRO in supplying materials, equipment and services and about 25 industries have begun using ISRO developed technology for non-space uses (Dhawan, 1983, p.15).

(*xi*) *Relations with User-Agencies*: As far as space applications are concerned, many user-agencies such as research organisations, autonomous bodies, government agencies and departments have a close link with ISRO. In this area, team effort from specialists of other agencies is also needed.

3.2.4 *ISAC Library*

It may be more appropriate, at this stage, to know in brief the background and status of the 'primary library'[7] used by the space technologists. The Library of ISAC came into existence in the last quarter of 1972 as a supporting facility to the then ISSP. Later, in 1976 when ISSP became ISAC, the Library attained the status of centre library. The library developed from a small collection of books and a few journals manned by non-professionals. However, the growth was very much accelerated after becoming the centre library and after induction of professional staff in 1978. ISAC library is a special library attached to the mission oriented organisation (*i.e.*, ISAC) and its scope encompasses a vast variety of heterogeneous subjects.

Table 3.1 presents the growth pattern of ISAC Library in terms of budget, staff, space, members, collection and services. The steep increase

TABLE 3.1: Growth of ISAC Library Budget, Staff, Space, Members, Collection and Service.

Year	Budget (Rs. in lakhs for F.Y.)	Staff			Space		Services*					
		Prof.	Non-Prof.	Total	Plinth Area (in sq.ft.)	No. of reading seats	Members	Reprographics (in 000's)	Issues (in 000's)	Consultations (in 000's)	Total use of documents (in 000's)	Collection
1972	1.02	0	1	1	800	16	NA	NA	NA	NA	NA	212
1973	1.20	0	1	1	800	16	200	NA	NA	NA	NA	1229
1974	1.53	0	2	2	800	16	263	NA	NA	NA	NA	1807
1975	1.80	0	2	2	800	16	285	NA	NA	NA	NA	2328
1976	3.40	0	3	3	800	16	315	NA	NA	NA	NA	3539
1977	5.20	0	4	4	1100	16	485	NA	NA	NA	NA	5005
1978	7.98	3	5	8	2500	30	633	135	NA	NA	NA	7274
1979	9.52	4	5	9	2500	20	692	185	7	NA	7	13306
1980	13.00	4	6	10	3000	10	783	233	15	31	46	29063
1981	20.00	4	6	10	4000	50	890	260	18	33	51	42753
1982	22.00	6	5	11	4000	40	982	300	23	47	70	54725
1983	28.00	9	8	17	4000	40	1160	300	32	64	96	65563
1984	27.00	10	8	18	8000	60	1210	400	25	64	89	79215
1985	24.60	11	8	19	8000	50	1305	505	31	91	122	90345

Key: NA – not available;

*, Other services are enumerated in Chapter 9 — use of library and user—interactions with the library;

F.Y. - (Financial Year).

Note : Budget excludes salary, office stationery, fixture, lighting and other utilities, import duties, etc.; Members include only those who have registered themselves and availed library borrower tickets.

TABLE 3.2 : Growth of ISAC Library Collection

Year	Current periodicals	Bound periodicals	Books	Reports	Standards	Microforms (Reports)	Trade catalogues	Reprints	Audio-visual materials	Total*
1972	0	0	212	0	0	0	0	0	0	212
1973	30	0	1229	0	0	0	0	0	0	1229
1974	60	0	1807	0	0	0	0	0	0	1809
1975	72	54	2220	54	0	0	0	0	0	2328
1976	96	405	2729	405	0	0	0	0	0	3539
1977	141	604	3761	604	0	36	0	0	0	5005
1978	208	749	5740	749	0	36	0	0	0	7274
1979	337	937	7639	1784	518	1898	304	197	29	13306
1980	533	1665	9040	2847	1355	13425	407	278	46	29063
1981	575	2224	9984	3369	2151	24163	484	316	62	42753
1982	645	2780	11462	4200	3065	31962	698	462	96	54725
1983	655	3943	12748	5011	3309	39022	800	630	100	65563
1984	631	4497	14158	5947	3991	48828	850	840	104	79215
1985	632	5670	15800	6832	4645	55512	890	891	105	90345

Key : *, The total excludes current periodicals.

after 1978 in almost all aspects is quite clear. Table 3.2 depicts growth of collection in terms of its components such as books, reports, journals, etc. The Library provides most of the routine services of a special library. Many services are quite informal. A large number of short-range reference queries are attended to by the library staff almost daily. The range of the services offered by the library include acquisition, reader assistance, indexing, lending, consultation, inter-libray loan, reprographic, micrographic, reference, translation, extension, documentation, current awareness, literature search, reader alert and abstracting services and regular as well as adhoc publications.

3.3 Demographic Characteristics

The demographic characteristics have a greater influence on a public library user than that of a special library user. However, three fundamental characteristics, namely, age, sex and experience have been considered in the present study.

3.3.1 *Age*

A noted feature of the space technologists is that they are quite young. The pattern of age distribution of the population and response population is shown in Table 3.3. It is evident that nearly two-third of the population is in the age group of 25 to 34 years, three quarter is less than 35 years old, nearly 96 per cent are below 40 years. The average age of an Indian space technologist is just 31 years as against 44 years and 40 years in case of ESA respondents of Raitt (1984, p. 121) and DOD user-study (Auerbach, 1965, p. B-3) respectively.

3.3.2 *Sex*

Research findings in mass communication suggest that ". . . there are consistent differences between men and women in their information-processing habits. Women seem to ingest, store and reproduce information with less distortion than do men" (Bauer, 1973, p.146). This aspect indicates that there is a greater need in library and information area to investigate these differences in information-behaviour of men and women. Females constitute 7.6 per cent and 6.5 per cent respectively of the population and response population. The percentage of females is in conformity with the overall employment pattern of females in R&D

TABLE 3.3 : Frequency Distribution of Age

Age in Year	Population No.	Population %	Response Population No.	Response Population %
Upto 24	95	11.6	69	13.2
25—29	268	32.6	173	33.0
30—34	255	31.0	165	31.5
35—39	151	18.4	85	16.2
40—44	39	4.7	27	5.1
45—49	6	0.7	1	0.2
50 and above	8	1.0	4	0.8
Total	822	100.0	524	100.0

Note: Excludes 15 not known cases under Population and 11 not answered cases under Response Population.

TABLE 3.4 : Frequency Distribution of Work Experience (as of 1983)

Experience in Years	Experience within ISRO				Experience prior to Joining ISRO		Total Experience	
	Population		Response Population		Response Population		Response Population	
	No.	%	No.	%	No.	%	No.	%
< 00	54	6.5	29	5.4	146	36.5	34	4.8
01—02	154	18.4	91	17.0	117	29.3	117	16.7
03—05	224	26.8	147	27.5	81	20.3	143	20.4
06—10	295	35.3	195	36.5	43	10.8	236	33.6
11—15	100	12.0	67	12.5	7	1.8	132	18.8
16—20	6	0.7	4	0.7	5	1.3	27	3.8
21—25	1	0.1	1	0.2	1	0.3	7	1.0
26—30	1	0.1	0	0.0	0	0.0	3	0.4
31—36	0	0.0	0	0.0	0	0.0	3	0.4
Total	835	99.9	534	99.8	400	100.3	702	99.9

Notes :

1. Experience in ISRO of two in the pupulation and one in the response population were not known/answered and hence excluded.
2. In case of experience prior to joining ISRO and total experience, there were 130 not answered/not known cases, 4 invalid responses and one error — all excluded from the table.

institutions in the country.[8] Females are mostly placed in the middle range in the status structure. As against this, Raitt (1984, p. 121) found only 2.8 per cent female respondents in his study of aerospace establishments. Though he has not analysed the population characteristics and sample characteristics, he concluded that "... aerospace is traditionally 'technical' and thus a man's world . . .", which is not supported by the present study.

3.3.3 *Length of Experience*

The data relating to length of experience within ISRO as well as prior to joining ISRO of the space technologists are collected through the questionnaire (Q.No.15.1). The length of experience of the respondents within ISRO is presented in Table 3.4 alongwith experience prior to joining ISRO. The table also shows the frequency distribution of experience within ISRO of the population. As shown in the Table 3.4 while only 6.5 per cent of the population is without any experience within ISRO, 36.5 per cent do not have any experience prior to joining ISRO. However, only 14 per cent have more than 5 years of experience prior to joing ISRO. Also nearly two-third of the respondents have either no experience or 1 or 2 years of experience prior to joining ISRO.

Little over 50 per cent of the population has less than 6 years of experience in ISRO. Whereas about 13 per cent of the population has more than 10 years of experience. The average experience of the space technologists within ISRO, prior to joining ISRO and the total experience are 5.9 years, 2.5 years and 7.4 years respectively. Further, the Table 3.4 makes it very clear that response population is highly representative of the total population in terms of experience.[9]

The frequency distribution of experience of the respondents of DOD User-study (Auerbach, 1965, p. B-6) was almost close to that of the present study. DOD study had 44 per cent respondents with over 5 years, 40 per cent with 1—5 years and 16 per cent with less than a year of experience.

Table 3.5 presents the responses to the question (No.15.2) about the nature of the organisation worked prior to the joining of ISRO. An important finding from the data in this Table is that nearly 41 per cent of the space technologists have prior experience from industries, followed by 30 per cent from R&D organisations and 18 per cent from academic/educational institutions. This very much substantiates the assessment of the nature of space technology and the space technologists presented previously that the nature of work is somewhere between R&D and industry (the assumption is that after changing the job, the majority continue to work on the same line). Raitt (1984, p.129) found almost an opposite trend that 49 per cent of his respondents in the aerospace establishments worked previously in academic environment and only 20 per cent worked in industrial environment though his classification was slightly different from that of the present study.

3.4 Personality-Characteristics (Work-Related)

Three work-related personality-characteristics[10], namely, motivation, job satisfaction and performance have been considered. There has already been controversies in the behavioural research about job satisfaction and performance. While some say satisfaction causes performance, the others say performance causes satisfaction and the third group says rewards (including promotion) cause both performance and satisfaction (Schermerhorn, 1982, p.63—68).

3.4.1 *Motivation (in Relation to Information-Seeking)*

The factors that motivate the respondents to seek information are discussed in detail in Chapter 4 (Motives and Purposes of Seeking Information). It is found that acquiring and updating knowledge in the field, self-improvement, achieving desired result in work, maintaining professional competence, pleasure of doing good work, self-fulfilment and self-satisfaction are the major and guiding motivators for the space technologists in seeking information. Some of the weak motivators are writing and publishing papers, having visibility among peers and colleagues, and having an edge over other competitors. Recognition, pursuing own research, continuing education, project reviews and departmental reviews (for promotion) are found to be the average motivators.

3.4.2 *Job Satisfaction*

Users were asked through the questionnaire (Q.No.16) the degree of job satisfaction[11] they have in their present jobs and the results are presented in Table 3.6. Surprisingly the nonresponse to this question is as low as 16 (out of 535). It is clear from the Table 3.6 that a majority (58.7%) of the respondents are fairly satisfied with their jobs.The fully satisfied and the fairly satisfied put together constitute nearly 73 per cent. The 'partially satisfied', 'not much satisfied' and 'not satisfied' are 16.1 per cent, 7.9 per cent and 3.1 per cent respectively. By considering the responses on a five-

TABLE 3.5 : Nature of Organisation Worked Prior to Joining ISRO

	Nature of organisation worked	No.	%
A	Academic/Educational	50	17.9
B	R & D	84	30.1
C	Industry	114	40.9
D	Others*	31	11.1
	Total	279	100.0

Key : *, 21 have specified as follows : Defence and Defence related organisations (5), Banks, Insurance Companies, P & T, P W D, Hotel, AIR, Private work and other State and Central Government Departments.

TABLE 3.6 : Job Satisfaction

	Job Satisfaction	No.	%
A	Fully satisfied	73	14.1
B	Fairly satisfied	303	58.7
C	Partially satisfied	83	16.1
D	Not much satisfied	41	7.9
E	Not satisfied	16	3.1
	Total	516	99.9

Note: Excludes two errors, one invalid response and 16 not answered cases.

point scale from 4 to 0, in the decreasing order of satisfaction, the weighted average satisfaction is 2.73, *i.e.*, close to or little less than fairly satisfied level.

3.4.3 *Performance*

It is very difficult to measure the 'performance' characteristic. By and large, there are four rough measures of performance. They are : (i) performance rating by immediate superiors as reflected in annual confidential reports, (ii) segregating high performers based on opinion of other colleagues *i.e.*, peers' judgement, (iii) the frequency of promotions secured, and (iv) performance based on professional activities and achievements such as lectures delivered, seminars attended, internal documents prepared, papers published, patents obtained, awards and citations received, etc. Quite often performance measures based on opinion are called 'subjective' measures and those based on concrete measures such as number of papers published are called 'objective' measures. Other measures such as 'productivity' and 'creativity' have also been used in some information-behaviour studies.

The first measure based on performance rating in annual confidential reports is a questionable one as each superior has his/her own way of rating. Further, it is not easily accessible due to its confidential nature.

3.4.3.1 ***Performance Based on Peers' Judgement:*** A secret opinion poll about high performers in the organisation was carried out at the time of interview requesting respondents to list the names of persons whom they think as high-performers in the centre. The final processed list of high-performers has contained 26 persons commonly cited/agreed by two or more respondents. It may be noted here that this method of peers' judgement is not totally free from bias and certainly the list is not exhaustive but only indicative. These 26 persons considered as high performers formed 3.2 per cent of the population and of them 20 persons who responded to questionnaire formed 4.0 per cent of response population.

3.4.3.2 ***Performance Based on Frequency of Promotion:*** The third measure was easy to ascertain but the variation in promotion policy at different times for different disciplines and at different levels in the hierarchy is its limitation.

The data about the number of promotions secured within ISRO by the space technologists has been elicited through the questionnaire (Q.No.21) and checked with records. The frequency distributions of number of promotions among the population as well as response population are shown in Table 3.7. The Table 3.7 shows that a little over one-third of the population has not yet secured promotion, another 30 per cent have secured one and the rest secured 2 to 4 promotions (except one who secured 5 promotions). It is clear from the table that the response population is representative of the population in terms of number of promotions secured though it has a slight under-representation of those having no promotion and slight over-representation of those having one promotion.

The data about the number of promotions secured together with data about length of experience in ISRO was used to arrive at the average number of years taken per promotion ignoring all extreme cases of over 12 years and below 3/4 years of experience. Since the normal time to get eligibility for a promotion is 3 years and 4 years for non-gazetted and gazetted ranks respectively, the same has been used as a threshold criterion to segregate high-performers from others. In this way 158 persons in the population are found to have taken, on an average, not more

TABLE 3.7 : Frequency Distribution of Number of Promotions

No. of promotions	Population		Response Population	
	No.	%	No.	%
0	270	34..3	147	28.1
1	238	30.2	183	34.9
2	151	19.2	109	20.8
3	102	12.9	71	13.5
4	26	3.3	14	2.7
5	1	0.1	0	0.0
Total	788	100.0	524	100.0

Note : Fortynine not known cases under population and 11 not answered cases under response population are excluded from the table.

than 3/4 years per promotion and they are called high-performers based on the frequency of promotion. Interestingly, this list included 16 of the 26 persons considered as high-performers based on peers' judgement (vide section 3.4.3.1). Some more cases would have been common to both the lists but for ignoring the extreme cases in the process of normalisation of average number of years per promotion. Out of 158 high-performers (who form 19.6% of the population, 100 have responded to the questionnaire forming 19.8% of the total respondents.

3.4.3.3 ***Performance Based on Professional Activities and Achievements***

The last measure of performance is based on professional characteristics discussed in detail in a later section (3.6.3). An attempt is made to develop professional activities and achievements index for each respondent with various quantifiable professional activities and achievements as its components. The index consists of three parts (Table 3.8). The first part includes number of lectures delivered, number of seminars, symposia and conferences attended and number of internal documents/reports prepared in last one year, average number of papers published average number of patents obtained and awards and citations received per year of experience (the latter two with a weightage of 5), number of professional associations, societies, board of studies, editorial committees in which one has a membership and number of professional journals personally subscribed. The second part consists of number of official tours during the last one year and number of committees in which one has membership. The third part consists of number of inservice trainings received so far and number of

TABLE 3.8 : Frequency Distribution of Professional Activities and Achievements Index

	Professional Activities and Achievements Index			
Frequency	**Part 1**	**Part 2**	**Part 3**	**Total composite Index**
00	63	205	172	26
01	27	43	100	19
02	27	36	22	12
03	28	20	5	9
04	19	21	5	15
05	29	13	1	16
06—10	42	33	0	45
11—15	28	12	0	15
16—20	3	9	0	15
21—30	7	3	0	14
31—40	2	0	0	8
41—50	0	0	0	2
Above 50	0	0	0	1
Total	275	395	305	197

Note: If data is not available for any one of the components, the case is dropped from the frequency distribution of total composite index.

part-time studies undertaken in last two years. In the development of the above index the final value has been rounded off to nearest integer and if any one of the components is not known the case has been excluded as not known

3.5 Organisation Oriented Environmental Characteristics

The term 'environment' is used here to refer to the 'organisational environment' prevailing in the ISRO. Such characteristics may also be called 'situational' characteristics. With respect to organisational environment the space technologists can be grouped in a number of ways. But selected attributes which are expected to have stronger relation with ISB are presented in this section. Since work-related information is considered in this study, the attributes arising out of the work situation are naturally very important.

3.5.1 *Status/Rank Based on Designation/Grade in the Organisational Hierarchy*

Though the criterion here is designation, it also represents the official grade/status and rank in the hierarchical structure of the organisation. Status appears to be the key characteristic for information-behaviour studies. Several studies have suggested that the existing information systems favour the older, higher-status scientists (National Research Council, 1967, p. 4-5). The organisation has 14-tier status structure which is roughly pyramidal in nature. Table 3.9 presents the status-wise distribution of, (i) the population, (ii) the response population, and (iii) the difference between the percentages of these two for each status. Nearly 80 per cent of the population is covered in the status groups E to J (Table 3.9). It may be noted that status groups A to H have only one designation in each group. But groups I to N have slightly different designations like scientific assistant, technical assistant, draughtsman, tradesman and laboratory assistant but of equal rank within each group. As status groups A,B,C,D,M and N have relatively less number of persons, for analysis purpose, groups A,B,C and D as well as M & N and sometimes K,L,M and N are treated together.

The status distribution of response population is more or less closely conforming to the status distribution of population. The only exception is found in status M where the difference in percentages of response population and population is -2.3, an under-response. Similarly, a slight over-response is found in case of status D and E. These differences in response rates further get reduced if we consider nine rejected responses and 14 persons to whom questionnaire could not be distributed. Hence response population does not cause any bias in the study.

3.5.2 *Technocrat-Nature Based on Functional Designation*

The roles with which users' perceive their information-requirements represent important characteristics. The roles such as manager, leader, teacher, guide, decision-maker, etc., are quite important in understanding information-behaviour. Some of the space technologists by their very functions and nature of work are given additional functional designations such as project director, deputy project director, project manager, deputy project manager, division head, section head, group head, coordinator, etc. Such designations of 80 (13.3%) persons out of 603 persons in the population as of January 1982 were noted in the user-record. In case of other 204 persons, it was not known whether they had any functional designations. Since the persons with functional designations have technical as well as managerial nature of work they are called technocrats. Out of 80 technocrats in the population, 48 have responded to the questionnaire

TABLE 3.9 : Status by Designation or Grade

Status		Total Population		Response Population		% of Response Population
Code	**Designation**	**No.**	**%**	**No.**	**%**	**-% of Population**
A.	Scientist/Engineer 'H'	1	0.1	0	0.0	– 0.1
B.	Scientist/Engineer 'G'	1	0.1	0	0.0	– 0.1
C.	Scientist/Engineer 'SG'	7	0.8	4	0.8	0.0
D.	Scientist/Engineer 'SF'	19	2.3	16	3.1	+ 0.8
E.	Scientist/Engineer 'SE'	74	9.0	52	9.9	+ 0.9
F.	Scientist/Engineer 'SD'	103	12.5	67	12.8	+ 0.3
G.	Scinetist/Engineer 'SC'	170	20.6	107	20.5	– 0.1
H.	Scientist/Engineer 'SB'	82	10.0	54	10.3	+ 0.3
I.	Scientific Assitant 'C' Technical Assistant 'C', Draughtsman 'E', Tradesman 'G' or equivalent	110	13.3	57	12.8	– 0.5
J.	Scientific Assistant 'B', Technical Assistant 'B', Draughtsman 'D', Tradesman 'F' or equivalent	104	12.6	68	13.0	+ 0.4
K.	Scientific Assistant 'A', Draughtsman 'C', Tradesman 'E', Senior Laboratory Assistant or equivalent	38	4.6	26	5.0	+ 0.4
L.	Draughtsman 'C2', Tradesman 'D', Laboratory Assistant 'C' or equivalent	63	7.6	42	8.0	+ 0.4
M.	Draughtsman 'A', Tradesman 'B', Laboratory Assistant 'B' or equivalent	41	5.0	14	2.7	– 2.3
N.	Tradesman 'A', Laboratory Assistant 'B', or equivalent	10	1.2	5	1.0	– 0.2
V.	Visiting Scientist	1	0.1	1	0.2	+ 0.1
	Total	824	99.8	523	100.1	

Note:

1. Table exclude one error and 12 not known cases in the Pupulation and 12 not answered cases in response population.
2. Nine questionnaire responses totally rejected are also excluded.
3. Fourteen persons included in the pupulation were not distributed the questionnaire.

accounting for 11.9 per cent of 402 respondents (excluding 133 respondents in whose case existence of functional designation was not known).

3.5.3 *Section, Division and Project Affiliation*

Table 3.10 and Appendix IV respectively depict the division/project-wise and section/project-wise distribution of the population. As noted earlier, division consists of one or more sections and such a division does not exist within a project. Hence, projects remain on par with both divisions and sections. Certain groups such as QA, PPED and TP which are 'deemed divisions' are covered as they are in a division-wise as well as a section-wise breakup. Further, some personnel are identified under both project and division/section. Here, the persons shown under projects are those who are exclusively working in project management offices or project core teams. It is clear that all divisions, projects and sections have almost been proportionately represented in the response population. A little over 5 per cent of the response population work exclusively in three major project management offices/project core teams. Four major technical divisions namely, AOCSD, IGTD, MSD and TTC&P account for 51 per cent of the response population.[12]

3.5.4 *Nature of Work*

The nature of work of an individual has a very strong influence on the work-related information-need and information-seeking and information-gathering behaviour. The nature of work of the space technologists has been ascertained through the questionnaire (Q.No.13.1). The responses are tabulateed in Table 3.11. Inspite of requesting the respondents to check any one type based on maximum time consumed in their day to day work, 75 respondents have checked more than one. Except a combination of Design and Development with Fabrication and Testing checked by 28 respondents (category P), all other combinations are invalidated. As revealed at the time of testing the questionnaire as well as noting responses, it was found extremely difficult to divide the nature of work into mutually exclusive groups.

The present grouping of the nature of work is in the descending order of the management/supervision content of the job and ascending order of the operational activity (except F-Others). It is clear from the Table 3.9 that nearly 68 per cent of the respondents are involved in design, development, fabrication and testing (C,P & D) activities. The management/

TABLE 3.10 : Division and Project Affiliation of the Space Technologists

Code	Acronym	Name	Population No.	Population %	Response Population No.	Response Population %
A	AOCSD	Attitude, Orbit Control and Sensors	86	10.4	60	11.5
B	EMD	Engineering Maintenance	23	2.8	15	2.9
C	FAC	Facilities	173	21.0	99	18.9
D	IGCTD	Integration, Ground Checkout and Testing	86	10.4	59	11.3
E	IRS	Indian Remote Sensing Satellite Project	21	2.5	15	2.9
F	MOPD	Mission Operations and Planning.	78	9.5	44	8.4
G	MSD	Mechanical Systems	88	10.7	45	8.6
H	PPED	Planning, Programming and Evaluation	17	2.1	12	2.3
I	INSAT II	Indian National Satellite II Project (Proto INSAT)	5	0.6	5	1.0
J	QA & TE	Quality Assurance and Technical Evaluation	59	7.2	39	7.5
K	SROSS	Stretched Rohini Satellite Series Project	16	1.9	8	1.5
L	TP	Technical Physics	19	2.3	19	3.6
M	TTC & P	Telementry, Tracking, Communication and Power	150	18.2	102	19.5
N	—	Others	3	0.4	1	0.2
		Total	824	100.0	533	100.1

Note: Table excludes 12 not known cases and one error under population and 12 not answered cases under response population.

supervision and planning/system analysis (A and B) are the nature of work of 20.5 per cent of the respondents as against 60 per cent supervisors found in Raitt's study[13] (1984, p.123).

Respondents were also asked through the questionnaire about any substantial change in the nature of their work in recent years (Q.No.13.2). Nearly 29 per cent (*i.e.*, 146 out of 505 valid responses) felt some change in the nature of their work in recent years. This is inspite of the fact that 120 (22.4%) were two or less than two years old in the organisation. The

TABLE 3.11 : Nature of work of The Space Technologists

	Nature of work	No.	%
A	Management/Supervision	42	8.7
B	Planning/System Analysis	57	11.8
C	Design and Development	162	33.4
P	C and D (Design and Development & Fabrication and Testing) @	28	5.8
D	Fabrication and Testing	140	28.9
E	Operational Activity	36	7.4
F	Others*	20	4.1
	Total	485	100.1

Key: @ Though respondents were requested to check any one type of nature of work based on maximum time consumed in their day to day work, as many as 75 have checked more than one. Except C and D combination all others were found to be checked by a maximum of 8 and hence invalidated. Sixteen have checked 3 or more categories and one has not answered; all these are excluded.

* Out of 20, two were close to D, Draughtsman Job-3, Key punching-2, Software Development-3, Inspection, Screening maintenance, servicing and calibration of electronic components-7, system studies-1, photography-1, coordination-1, and project core team-1.

effect of such a change in nature of work on ISB is discussed in a later chapter.

3.5.5 *Major Specialised Field of Activity*

Respondents were requested to mention the major specialised field of activity in their own words (Q.No.13.3). Responses to this open ended question varied quite widely. Seventy eight respondents have not answered and another 100 are invalidated for vague answers.[14] The rest of 346 have mentioned fairly specialised activities. Fields were sequentially coded and related activities grouped together as questionnaires were received and processed. It resulted in 81 different fields in the first stage indicating the professional diversity of the space technologists.[15] With the help of subject experts these fields have been regrouped into 19 main fields (broad level) as shown in Table 3.12 and 81 sub-fields as per Appendix-5. It is clear from the table that largest number of respondents are working in the area of mechanical engineering, communication engineering, computer science, instrumentation and physics. Inspite of wide variations between the fields of activities of the respondents of DOD user-study (Auerbach, 1965, p. B-8) and the present study, communications and electronics with 15 per cent respondents and physics with 6 per cent respondents scored highest in DOD user-study and it is quite close to the composition in the present study (see Table 3.12).

3.5.6 *ISAC Library-Oriented Characteristics*

Various interactions with the library including use of the library by the IST is discussed in a later chapter (Chapter 9). As a background data about

TABLE 3.12 : Major Specialised Fields of Activity of the Space Technologists

	Major Specialised Fields of Activity	No.	%
A	Physics	26	7.3
B	Earth Observations	7	2.0
C	Instrumentation	35	9.8
D	Control Engineering	11	3.1
E	Structural Engineering	22	6.2
F	Reliability and Quality Engineering	16	4.5
G	Power System	20	5.6
H	Mechanical Engineering	57	16.0
I	Electronics	9	2.5
J	Electronic Packaging	20	5.6
K	Microprocessor Based Systems	8	2.2
L	Communication Engineering	40	11.2
M	Space Simulation	17	4.8
N	Propulsion	1	0.3
Z	Computer Science	40	11.2
P	Management	17	4.8
Q	Safety Engineering	1	0.3
R	Engineering Maintenance	5	1.4
S	Photography	4	1.1
	Total	356	99.9

Note: 1. Excludes 100 invalidated responses, 78 not answered and one error

2. Appendix VIII depicts the specific fields as expressed by respondents grouped under each of these categories with appropriate score.

four aspects relating to ISAC Library, namely, Library membership, Library Committee membership, authorisation for xerox services of the library and participation in an voluntary abstracting service are presented in this section.

First, 4.7 per cent or 38 space technologists in the population, who had the eligibility to become members of the primary (ISAC) library, were not members at the time of this study. It may also be noted that without being a member one can make 'inhouse' use of the library. Interestingly, the non-members are found in almost all the status groups from E to N. Out of the 38 non-members, 16 have responded to the questionnaire constituting 3.2 per cent of the respondents.

Secondly, 24 (3%) space technologists have served, at least once, as members in the library committee during 1973 and 1983 and participated in collection development and framing the policies of the library. However, the library committee had no member below the rank (status) F and a large majority of the members were of either D or E status. Out of 24 members/ ex-members of library committee, 18 persons have responded to the questionnaire accounting for 3.6 per cent of the respondents.

Thirdly, the space technologists who had authorisation to approve and directly avail reprographic service of the library constituted 6.8 per cent of the population (55 persons) of the rank (status) A to G. The response population contained 6.4 per cent (32 persons) of such persons with power to approve reprographic requests.

Lastly, 13.9 per cent (112 persons) of the population have volunteered for Satellite Technology Abstractors Group (STAG), a local abstracting service of the ISAC library. However, only 2 per cent (17 persons) are found active in the group. Interestingly, though the volunteers for this group are from status C to J, only persons at the middle rank *i.e.,* D to G have actively contributed to the service. The response population included 79 (or 15.7%) such volunteers of STAG.

3.6 *Professional Characteristics*

Various professional qualifications, qualities, activities and achievements of the space technologists which have a bearing on the areas of their interest are discussed in this section. They are not necessarily the results of work situation, but they are closely related to work environment in the organisation.

3.6.1 *Educational Qualifications*

In the context of information-preferences in mass communication Bauer (1973, p.145) says "probably the most generally recognised relevant personal characteristic is education and/or intelligence". The highest educational qualification attained by the space technologists was elicited through questionnaire (Q.No.14.1) and checked with available data from records. Wherever more than one qualification of the same level is cited, the latest-acquired is considered. The data regrouped for both population and response population at the broad level such as undergraduates, diploma holders, graduates, post-graduates and doctorates is shown in Table 3.13. Interestingly, the response population is highly representative of the population in terms of educational qualifications.

The Table 3.13 shows that about one-fifth of the population consists of undergraduates and another one-fifth diploma holders. Graduates and post-graduates constitute 27 per cent each. The rest of about 5 per cent are doctorates. While the percentage of graduates and post-graduates are close to the average level in R & D institutions in the country, the diploma holders and undergraduates are higher than the country's average and doctorates are less than half of the country's average[16]. The DOD user-study (Auerbach, 1965, p.B-4) had 8 per cent doctorates, 17 per cent postgraduates, 65 per cent graduates and others constituted rest of 10 per cent.

Further, the response population is slightly under-represented by undergraduates and diploma holders and over-represented by graduates

TABLE 3.13 : Qualifications of the Space Technologists

Qualifications	Population		Response Population	
	No.	%	No.	%
Undergraduates	145	20.2	96	18.5
Diploma holders	148	20.7	98	18.9
Graduates	196	27.3	153	29.5
Postgraduates	196	27.3	149	28.7
Doctorates	32	4.5	23	4.4
Total	717	100.0	519	100.0

Note: 1. The Table excludes 120 not known cases under population and 16 not answered cases under response population.

2. See Appendix IX for explanation of codes of qualifications.

and post-graduates compared to the (structure of) population.

The pattern of educational qualifications of the space technologists are shown in Appendix-VI. The significant types of qualifications are: undergraduate certificates such as ITI, NAC and NTC (116), diploma in electronics, communication, telecommunication and radio engineering (72); diploma in mechanical engineering (54); degrees like BE (123), B.Sc. (22), B.Tech. (16), ME(45), M.Sc. (63), M.Tech. (64), and Ph.D. (32).

3.6.2 *Subject of Specialisation*

Respondents were requested to check their broad area of specialisation in the highest qualification from a list of subjects. The data (excluding 18 responses invalidated for checking more than one subject, 7 not answered and 3 errors) is presented in Table 3.14. Also presented in the table is approximate distribution of subject of specialisation among population as of January 1982 from earlier records.

It is clear from the Table 3.14 that response-population is quite representative of the population. Electronics is the specialisation of largest number (44%) of the respondents, which is followed by mechanical engineering (27%) and Physics (8%).[17] Interestingly, Raitt (1984, p.127) also had the highest percentage of the respondents with specialisation in physics, electronics, aeronautical engineering and electrical engineering.[18]

Table 3.15 presents the pattern of broad disciplines of the IST such as pure sciences, applied sciences/engineering/technology and management based on the specialisation as well as qualifications of the space technologists (doctorates and vocationally trained personnel have been categorised separately in this Table 3.15). As per this Table, nearly 62 per cent of the population belongs to applied science, engineering and technology, about 13 per cent to pure science and only 1.3 per cent to management. The response population is also found to be highly representative of the population on this count too.

3.6.3 *Other Professional Activities and Achievements*

There are many professional activities and achievements of the Indian space technologists which enhance their contact with information-sources and result in increased information-transfer. These were ascertained through the questionnaire (Q.No.12). The frequency distribution of the

TABLE 3.14 : Subject of Specialisation of the Space Technology

	Specialisation in Highest Qualification	Population (as of Jan 82)		Response Population	
		No.	%	No.	%
A	Physics	60	8.3	42	8.3
B	Chemistry	4	0.6	4	0.8
C	Mathematics	(+J)36	5.0	20	3.9
D	Civil Engineering	3	0.4	3	0.6
E	Mechanical Engineering	233	32.1	137	27.0
F	Aeronautical and Structural Engineering			19	3.7
G	Thermal Engineering			7	1.4
H	Electrical Engineering	16	2.2	17	3.4
I	Electronics	317	43.7	222	43.8
J	Computer Science	(See C)		9	1.8
K	Chemical Engineering/ Technology			1	0.2
L	Material Science			4	0.8
M	Management			9	1.8
N	Others*	57	7.9	13	2.6
	Total	726	100.2	507	100.1

Key: *, Others included Metallurgy, Botany, Controls and Dynamics, Statistics, Optics, Atmosphere, Science, Electronics-Instrumentation, Photography, Space Dynamics, Keypunching, etc.

Note: Eighteen responses invalidated, 7 not answered and 3 errors are not shown in the table.

data relating to tours, lectures, seminars, internal documents prepared, papers published, patents obtained, awards and citations received, memberships in professional bodies, committees, board of studies and editorial committees, professional journals personally subscribed, part-time studies undertaken and in-service trainings received are presented in Tables 3.16A and 3.16 B. These activities and achievements of the space technologists have been used as components in the development of professional activities and achievements index discussed earlier in Section 3.4.3.3.

3.6.3.1 ***Tours, Lectures and Seminars***: The first three columns in Table 3.16A show the distribution of the number of lectures delivered,

TABLE 3.15 : Subject of Specialisation and Qualifications

	Population		Response Population	
	No.	%	No.	%
Doctorates	32	4.5	23	4.4
Pure Sciences	85	11.9	66	12.7
Applied Science, Engineering and Technolgoy	443	61.8	326	62.8
Management	9	1.3	7	1.3
Vocationally Trained Undergraduates	145	20.2	96	18.5
Others	3	0.4	1	0.2
Total	717	100.1	519	99.9

Notes: 1. Excludes 120 not known cases under population and 16 not answered cases under response pupulation.

2. See Appendix-IX for explanation of codes of qualifications.

number of seminars, symposia and conferences attended and the number of times other centres and organisations officially visited (*i.e.*, number of tours) during the last one year. The Table shows that nearly 69 per cent of the population did not have any lecturing or tour programme during a year. Similarly 51 per cent did not participate in any seminar as against 24 per cent (in last two years) and 51 per cent, respectively, found by Raitt (1984, p.188) and Shuchman (1981, p.40). Those who had lecturing and seminar participation had frequencies upto 10 (except two each in the range of 15-20) in a year. However, the number of tours of some has gone beyond 10 in a year and reaches the range of 21-30 in a year indicating thereby an intensive touring by a few.[19]

3.6.3.2 *Internal Documents (Reports) Prepared and Papers Published:* The publication rate has been considered as one of the important characteristics of scientists and academicians. The next two columns in Table 3.16A give the frequency distribution of the number of internal documents/reports prepared in a year and number of papers published so far and these activities are not done by 43 per cent and 68 per cent of the population, respectively, as against 29 per cent and 43 per cent (in last two years) found by Raitt (1984, p. 185).[20] The total number of internal documents prepared is 804 in a year and, hence, the average number of documents prepared per respondent per year is 1.9 (1.8 in case of Raitt's

TABLE 3.16A : Frequency Distribution of Professional Activities and Achievements (Percentages in bracket)

Freqency	Lectures delivered in a year	Seminars attended in a year	Tours in a year	Internal documents prepared in a year	Papers published so far	Patents obtained received so far	Awards and citations received so far	Membership of professional bodies Questionnaire from response records	
00	283 (68.7)	216 (50.8)	228 (68.6)	182 (42.9)	285 (68.2)	387 (98.5)	341 (86.3)	288 (70.2)	473 (84.0)
01	60	77	56	63	37	6	45	73	73
02	35	65	51	61	17	0	7	32	6
03	13	30	21	42	12	0	2	14	8
04	9	20	12	27	11	0	0	3	3
05	4	8	19	21	6	0	0	0	0
06	0	4	10	9	7	0			
07	1	0	3	3	4	0			
08	2	1	1	2	4	0			
09	0	0	1	2	0	0			
10	3	2	9	8	5	0			
11	0	0	0	0	1				
12	0	0	2	1	2				
13	0	0	0	0	2				
14	0	0	0	0	1				
15	1	2	3	1	11				
16—20	1	0	3	1	4				
21—30	0	0	1	0	5				
31—40	0	0	0	0	2				
41—50	0	0	0	1	2				
Sub-Total	412	425	420	424	418	393	395	410	563
Invalid	2	2	0	2	0	0	0	1	NA
Not answered	121	108	115	109	117	142	140	124	244
Total	535	535	535	535	535	535	535	535	807

Key: NA = Not Applicable.

TABLE 3.16 B: Frequency Distribution of Professional Activities and Achievements (Percentages in bracket)

Frequency	Membership in committees	Membership in board of academic institutions	Membership in editorial committees of journals	Professional journals subscribed	Part-time studies in last two years	Inservice training received (from records)
00	294 (73.7)	335 (85.2)	374 (95.9)	315 (80.2)	264 (65.3)	487 (86.0)
01	33	44	10	44	111	64
02	25	10	6	25	22	7
03	13	4	0	7	1	6
04	10	0	0	2	6	2
05	6	0	0	0	0	0
Sub Total	399	393	390	393	404	566
Invalid	1	1	1	1	1	NA
Not answered	135	141	144	141	130	241
Total	535	535	535	535	535	807

Key : NA = Not Applicable.

Note : Out of 139 who undertook some part time studies, 68 have specified the courses—four Ph.Ds, one each M.S. and M.Sc (Engineering), 34 were doing BE, AMIE and GRAD, IETE, 17 Diploma in Engineering including computer engineering, 9 short-term profeciency courses.

study). Similarly the total number of papers published by the respondents is 922 and hence the average is 2.2. (1 in case of Raitt's study).[21]

3.6.3.3 ***Patents, Awards and Citations Received:*** Table 3.16A (columns 7 and 8) gives the data about patents and awards received so far by the respondents. Only 6 persons (or 1.5% of the respondents) obtained patents, 54 persons (or 3.7%) received awards and citations in recognition of their professional contribution. These two criteria carry a weightage of 5 in professional activities and achievements index.[22]

3.6.3.4 ***Membership in Professional Bodies:*** The last two columns of the Table 3.16A give the frequency distribution of membership in professional associations, societies, etc. of the IST as per response to questionnaire as well as from the available records. Since, the data obtained from the records are older by more than a year and since new entrants also provided the data through questionnaire, the table shows less number of membership in case of data from records. However, patterns of both frequencies are quite close. About 30 per cent of the respondents have membership with atleast one of the professional bodies. The maximum number of memberships any respondent has is with four professional bodies, and the average works out to be 1.6 per respondent.[23]

3.6.3.5 ***Membership in Committees:*** The Space technologists, as a part of their work, play a subsidiary role of working on many committees such as purchase committees, recruitment committees, reviewing committees (for promotion), library and seminar committee and many other committees of technical as well as semi-technical nature. Participation in such committees is expected to increase the opportunity for inter-personal communication for members. The respondents were requested to mention the number of committees (both within and outside the ISRO) in which they are members or chairmen. As shown in Table 3.16B, the number of committees served by the respondents ranged from 1 to 10 (except one person serving 14 committees). Little more than 24 per cent of the respondents were members in atleast one of the committees and the average number of committees represented by the space technologists is 3.3.[24]

3.6.3.6 ***Membership in Board of Studies of Academic Institutions and in Editorial Committees of Professional Journals:*** Fifty-eight (15%) respondents as shown in Table 3.16 B and sixteen (4.1) are on editorial committees of Professional Journals and various boards of studies of academic institutions.[25]

3.6.3.7 ***Professional Journals Personally Subscribed:*** The respondents were asked through the questionnaire the number of professional journals to which they subscribed apart from those journals received as gifts. The responses tabulated in Table 3.16B show that as many as 78 (20%) have subscribed to atleast one journal. These journals naturally form part of the personal collection of the space technologists and the dependence on and role of the personal collection is explored in a later chapter.

3.6.3.8 ***Part-Time Studies Undertaken:*** The space technologists who have undertaken some part-time studies for certificate, diploma or degree, during the last two years were identified through the questionnaire and the data is tabulated in Table 3.16B. Thirty five percent of the respondents are involved in some sort of part-time studies, the duration and depth of which varied from short term proficiency courses to Ph.D.[26] As mentioned in the note to the table, 68 out of 139 have specified the courses. Half of those who disclosed this information are involved in B.E., AMIE and GRAD IETE and one-fourth in diploma in various engineering subjects.

3.6.3.9 ***Inservice Trainings Received:*** Lastly, the data about inservice trainings received by the space technologists as per the personnel information records, is presented in Table 3.16B. As many as 79 persons (or 14% of the population) have one to four inservice trainings during their stay at ISRO.

3.7 Inter-Correlation/Inter-Association of User-Characteristics

Though inter-correlation of user-characteristics is not the purpose of this study, an attempt is made in this section to examine the inter-correlation or inter-association among selected user-characteristics so as to choose a manageable size of 5-6 significant characteristics for further correlational analysis with information-behaviour. Based on the preliminary observations of the data, user-characteristics have been classified into five groups. It has been hypothesised that characteristics within each group are highly dependent and hence correlated/ associated. However, it does not imply that there is no relation between characteristics form different groups. If a significant high correlation/association exists between two characteristics of a group and one of them has been selected for further correlation with information-behaviour, then the findings can be extended to the other characteristics of the group (by transitivity property) as the data for both correlation/association tests have been drawn from the same population. For example, if age and experience are significantly and highly correlated in Group I, and if experience is chosen for further correlation study with information-behaviour, then the findings in respect of experience can be extended, to some extent, to the characteristic 'age' of the respondents.

3.7.1 *Group I : Age, Experience and Status*

A high rank order correlation is found between mean age and status (r_s = 0.7333, $p<0.02$, $df=8$) as well as mean experience and status ($r_s=0.6424$, $p<0.05$, $df=8$) of the respondents. A very high and significant correlation ($r=0.8167$, $p<0.001$, $df=390$) is found between age and experience of the respondents.

3.7.2 *Group II : Nature of Work, Technocrat Nature and Status*

A very high association ($x^2 = 1294.6, p < 0.001, df = 3$) is found between technocrat nature and status. Similarly the five types of nature of work, namely, Management/Supervision (A), System Analysis/Planning (B), Design and Development (C), Fabrication and Testing (D) and Operational Activity (E), associated statistically significantly ($x^2 = 223.2627$, $p < 0.001$, $df = 12$) with status of the respondents. The same chi-square test has shown a very high association ($x^2 = 75.0093$, $p < 0.001$, $df = 5$) between technocrat nature and nature of work of the respondents.

3.7.3 *Group III : Qualifications and Status*

There is a clear association of qualifications and status of the respondents with 99.9 per cent significance as per chi-square test ($x^2 = 558.4618$, $p < 0.001$, $df = 12$). There is a linear statistically significant association ($x^2 = 75.8052$, $p < 0.001$, $df = 20$) between age and qualifications of the respondents (the average age of doctorates is highest).

There is a curvilinear significant association ($x^2 = 35.7051$, $p < 0.05$, $df = 20$) between experience and qualifications of the respondents with undergraduates and doctorates having the highest average experience and graduates the lowest.

3.7.4 *Group IV : Subject of Specialisation, Major Field of Activity and Section/Division and Project Affiliation*

The chi-square test showed significant association between specialisation and field of activity ($x^2 = 119.4919, p < 0.001, df = 9$), specialisation and section/division and project affiliation ($x^2 = 39.336$, $p < 0.001$, $df = 6$), as well as field of activity and section/division and project affiliation ($x^2 = 76.7954$ $p < 0.001$, $df = 6$) of the IST.

3.7.5 *Group V : Job Satisfaction, Performance based on Frequency of Promotion and Performance based on Professional Activities and Achievements*

Research in management and behavioural science has shown that job satisfaction is positively related to productivity (*i.e.*, performance), level of occupation (*i.e.*, status), age and experience. Further, employees with low job satisfaction are likely to leave their employers (Davis, 1977, p.75-77). A non-parametric chisquare statistic test showed a statistically significant association of degree of job satisfaction with higher performance based on frequency of promotion ($x^2 = 9.9708, p < 0.05, df = 4$). But the same test showed no significant association ($x^2 = 26.4149, p > 0.05, df = 32$) of degree of job satisfaction with performance based on professional activities and achievements (index) which is contradictory to the findings of Vincent and Mirakhor (1972, p.51).

Surprisingly, there is no significant association of job satisfaction with status, (contradictory to that found by Meltzer and Slater, 1962, p.360), age, experience as well as the respondents leaving the organisation.

The performance based on professional activities and achievements of the respondents showed slight positive correlation with age ($r = 0.12$), experience ($r = 0.15$) and the number of promotions secured ($r = 0.12$) and supported the findings of Meltzer (1956, p. 32). The chi-square statistic test showed no association of performance based on professional activities and achievements with sex and job satisfaction, but significant association with qualifications ($x^2 = 175.6591, p < 0.001, df = 32$), subject of specialisation ($x^2 = 78.6960, p < 0.001, df = 40$), and the nature of work ($x^2 = 114.2934, p < 0.001, df = 48$).

Further, status is found to have a very high ($r_s = 0.9394, p < 0.001, df = 8$) correlation with performance based on professional activities and achievements (index) as against a curvilinear relation found by Meltzer and Slater (1962, p.360). Even two of the three components of the index showed the same level of correlation (r_s is 0.9879, 0.9879, 0.1788 respectively for the three components). A chi-square test between status and performance based on frequency of promotion showed no significant association.[27]

The above inter-correlations/inter-associations of user-characteristics indicate that the characteristics like status, nature of work, qualifications, subject of specialisation and performance (based on professional activities and achievements index) of respondents can quite fairly represent user-characteristics for a correlational study of information-behaviour of the space technologists.

NOTES

1. As per the definition of UNESCO, India has 23, 28, 200 scientists, engineers and technicians as of 1977 and ranks eighth (UNESCO, 1983, p, v-24).
2. The R & D expenditure of major agencies during 1983-84 as per Planning Commission (1984) was Rs.793 crores [India, DST, 1985, p (ii)] The expenditure of DoS during 1983-84 was Rs.166.55 crores. Hence the DoS expenditure is 21 per cent of total expenditure of major agencies. However, by ignoring unusual onetime expenditure of Rs.59.83 crores on INSAT-I during 1983-84, the percentage gets reduced to 13.5.

3. The data in this chapter has been drawn from the Annual Reports of DOS and ISAC, the Performance Budget Reports of DOS, informal sources within the organisation.
4. An examination of distribution of funds among various centres, schemes and units for the years 1981-82, 82-83 and 83-84 revealed that ISAC's share is 11.1, 14.6 and 13.8 per cent respectively.
5. Group D is the lowest of the cadre of groups A,B,C and D as per Government of India classification of employees.
6. In October 1984, divisions have been redesignated as groups, and sections as divisions.
7. "The concept of a 'primary library' is based upon the premises that every user should have atleast one library that he can make use of which will put him in touch with the library system" (Hoadley and Clark, 1972, p.133).
8. "Out of 1.84 lakh personnel employed in R & D institutions only about 6 per cent were reported to be females" (India, DST, 1982, p.5).
9. The experience distribution against status (grade) revealed that the lower-level staff (status J to N) have 10 or less than 10 years of experience and higher-level staff (Status A to E) have 5 or more years of experience within ISRO. The experience prior to joining ISRO is distributed almost uniformly with all levels (Status) except the lowest two grades (M & N) (See Table 3.9 for more details about status).
10. "Personality is specifically used to represent the overall profile or combination of traits that characterise the unique nature of a person" (Schermerhorn, 1982, p.107). Though many personality traits could not be considered, the term is used in a limited sense to represent three work-related personality characteristics.
11. The job satisfaction is defined as the degree to which one's job satisfies one's needs or the favourableness or unfavourableness with which an employee views his work.
12. A cross tabulation of division/project against status showed that EMD and FAC have more lower-level and less higher-level personnel than other divisions and projects. Obviously the same holds good for the sections within these divisions.
13. As admitted by Raitt (1984, p.123), there appears to be some biases caused in his study due ambiguity in the question and self projections of those who were 'in-charge' of some tasks.
14. Among vague answers, 28 responses are of the type 'mechanical', 'mechanical engineering', 'mechanical fields', and 'mechanical-related', and another 42 are with similar responses related to 'electronics'. Nine vague statements are related to the nature of work or nature of organisation/activity or qualifications. Not surprisingly, a large majority of 'not answered' and 'invalid' responses are from middle and lower middle levels (status). Ten persons have mentioned two or three heterogeneous activities. In these cases the first mentioned broadest activity has been considered.
15. It may be noted here that the major specialised field of activity obviously overlaps with the nature of work and division/project/section affiliations previously discussed as well as subject of specialisation in highest qualification to be discussed in the subsequent section.
16. "Out of 64,875 personnel directly engaged in R & D, by level of qualifications, 11 per cent were Ph.Ds., 31 per cent were post- graduates, 30 per cent graduates, 13 per cent diploma holders and the rest 15 per cent had other qualifications" (India, DST, 1982, p. 5).
17. For correlating the ISB data against specialisation in later chapters the specialisations A,C,E,F, H and I in Table 3.14 with more than 10 respondents have only been considered.
18. When specialisation of the respondents was examined against their status it is found that no specialisation except electronics and mechanical engineering has persons lower

than grade J. In other words, most of the lower cadre persons have either electronics or mechanical engineering as their specialisation.

19. These activities are evidenced more with higher-level personnel. However, lecturing and attending seminars are spread upto status L, and touring activity is restricted upto status K.

20. The percentages of engineers who presented seminar papers, published articles and wrote books in Shuchman's study (1981, p.43). are respectively 31,25 and 1.

21. Again these activities are also more in higher-level persons and, by and large, persons of status not lower than 'I' have participated in it (except participation of 9 persons in internal document preparation). A detailed discussion about publishing pattern of the space technologists is presented in Chapter 8.

22. No person lower than status J has obtained patents and similarly no person lower than K has received any award.

23. The membership with professional bodies is also more at higher-status and except three persons of status K and four persons of status L, no person lower than status J has membership with professional bodies.

24. By and large, the committee membership is restricted to scientists and engineers in terms of designation (*i.e.*, upto status H).

25. These persons are not below status J (except one of status K). Sixteen (or 4%) respondents of not below the rank of J are on editorial committees of one or two professional journals as shown in the Table 3.16.

26. These persons belonged to status groups D to M.

27. It may be noted here that in all the above calculation of chi-square statistic, attribute types having very few respondents, have been either eliminated or merged with closely related attribute type so as to avoid fallacy of chi-square test where more than 20% of the cells of expected frequency table have 5 or less values.

4

MOTIVES AND PURPOSES OF SEEKING INFORMATION OF THE INDIAN SPACE TECHNOLOGISTS

4.1 Motives of Seeking Information

The motivation of the space technologists for seeking information has been ascertained in the present study through the questionnaire (Q.No.11) by requesting them to check on a five-point scale certain identified list of motives as noted in Table 4.1. The motives enlisted in the questionnaire are based on the results of preliminary discussions with selected representative users and informal observations. These motives are present in different degrees with many users in the population. Hence, how they are ranked, how they are related to ISB, how strongly they are felt and finally, how they are related to selected characteristics of users are discussed here. It is clear from Table 4.1 that 34.2 per cent of the respondents considered writing and publishing (K) as a clear non-motivator. Other significant non-motivators are : have an edge over competitors (C) checked by 25.4 per cent, pursuing continuing education (L) checked by 23.2 per cent, and gaining recognition (D) checked by 20.6 per cent. On the other hand, acquiring and updating knowledge in the field (H) is considered a motivator by the largest number of the respondents (99 per cent) followed by self improvement (G) (98.6 per cent), achieving desired result in work (J) (98.5 per cent), maintaining professional competence (I) (97.7 per cent) and pleasure of doing good work, self-fulfilment and self- satisfaction (M) (95.9 per cent). Figure 4.1 pictorially depicts the strength of various motives of seeking information of the IST.

Based on the weighted mean score shown on Table 4.1 the motives of seeking information of the IST are ranked below:

TABLE 4.1 : Motives of Seeking Information

	Motives	Non Motivator (0)		Motivator				Total		Mean	S.D
		No.	%	1	2	3	4	No.	%		
A.	To pursue own research related to work/in the field	75	18.1	36	111	96	97	340	81.9	2.25	1.39
B.	To have visibility among peers and colleagues	79	19.3	75	144	95	17	331	80.7	1.75	1.13
C.	To have an edge over other competitors	105	25.4	57	114	105	33	309	74.6	1.77	1.29
D..	For recognition	87	20.6	63	118	108	46	335	79.4	1.91	1.29
E.	To prepare for departmental review	54	12.3	48	155	119	63	385	87.7	2.20	1.19
F.	To prepare for project review	70	16.1	45	159	109	51	364	83.9	2.06	1.21
G.	For self improvement	7	1.4	6	76	181	217	480	98.6	3.22	0.86
H.	To acquire and update knowledge in the field	5	1.0	10	81	184	206	481	99.0	3.19	0.84
I.	To maintain professional competence	11	2.3	14	101	165	181	461	97.7	3.04	0.96
J.	To achieve desired result in work	7	1.5	14	80	178	200	472	98.5	3.15	0.89
K.	To write and publish	153	34.2	92	130	58	15	295	65.8	1.31	1.16
L.	To persue continuing education	103	23.2	56	96	117	72	341	76.8	2.00	1.40
M.	For pleasure of doing good work, self-fulfilment and self-satisfaction	18	4.1	17	75	174	151	417	95.9	2.97	1.03
N.	Other motivational factors	138	83.1	7	7	6	8	28	16.9	0.43	1.06

Key : O, Non-motivator; 1, Weakest motivator; 2, Average Motivator ; 3, Fairly strong motivator; 4, Strongest motivator ; S.D, Standard Deviation.

Note : 1. Three hundred and sixty four respondents have left 'Other motivational factors' (N) unanswered.

2. Not answered respondents varied from 34 to 123 and invalid responses were upto 13 and both are excluded from the table.

3. The mean score for ananymous responses worked out to 1.83 , 1.21, 0.54, 1.28, 2.15, 1.62, 3.15, 3.15, 2.88, 3.38, 1.52, 2.12, 2.96 and 0.70 respectively for categories A to N.

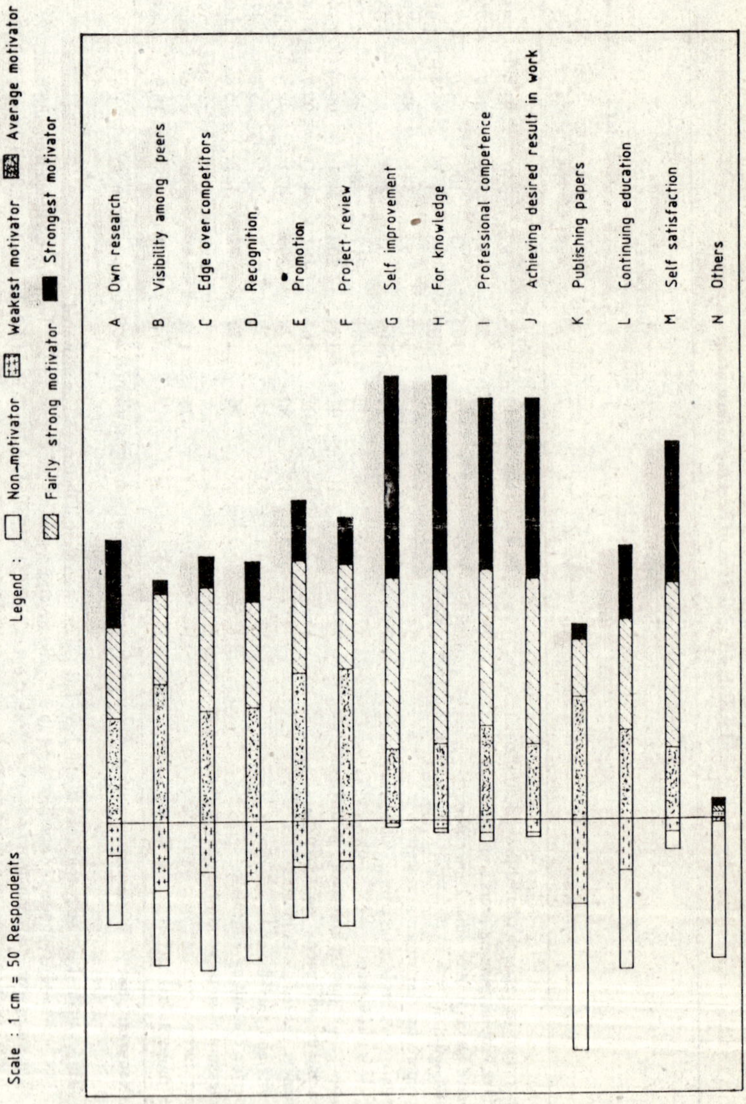

Fig. 4.1 : Motives behind seeking information

Rank	Code	Motives
1	G	For self-improvement
2	H	To acquire and update knowledge in the field
3	J	To achieve desired result in the work
4	I	To maintain professional competence
5	M	For pleasure of doing good work, self-fulfilment and self-satisfaction
6	A	To pursue own research related to work/in the field
7	E	To prepare for departmental review for promotion
8	F	To prepare for project review
9	L	To pursue continuing education[1]
10	D	To seek recognition
11	C	To have an edge over other competitors
12	B	To have visibility among peers and colleagues
13	K	To write and publish

Further by adding/subtracting one standard deviation to / from mean, it is found that 68.3 per cent of the respondents have checked the motivators G, H, I and J in the range of more than 2.36, 2.35, 2.08 and 2.26 on a five-point scale respectively. Self improvement (G), acquiring and updating knowledge in the field (H), achieving desired result in the work (J) and maintaining professional competence (I) being the strongest motivators occupy predominant position among the motives of seeking information. Shuchman (1981, p. 29) also made a similar observation about American industrial engineers when he said "most professionals are concerned with appearing technically competent to their colleagues" though the style differed. It appears respondents have placed 'self progress' a little higher than 'work progress' in their motives. A typical respondent writes "I gather information mainly to imporve my knowledge . . . I want to spend my leisure time for information gathering". In fact a couple of respondents have expressed that their information-seeking activity is mainly the leisure-time activity. This is roughly in conformity with Menzel's (Columbia University, 1958) observation that achievement, curiosity, self-evaluation and affiliation are the main motives of seeking information for scientists.

The next in the rank order are the three average motivators (*i.e.,* self fulfilment / self-satisfaction (M), pursuing own research (A), and preparing for departmental review for promotion (E)) which revolve round the ego of the seeker of information. It is very clear from the above data that the urge to write and publish which is one of the predominant motives of scientists in general, takes the last place in priorities among the space technologists. Similarly, the motives - recognition, visibility among peers and having edge over competitors—also score low ranking.[2]

4.2 Correlation of Motives of Seeking Information with (Selected) User-characteristics

Table 4.2 presents the results of association/correlation tests of motives of seeking information with six characteristics, namely, status, qualifications, nature of work, subject of specialisation, length of experience and professional activities and achievements of the IST. The distributions of mean score of the responses for each motive against status, qualifications, nature of work and specialisation are respectively shown in the Tables 4.3, 4.4, 4.5 and 4.6.

It is clear from Table 4.2 that motives of space technologists in seeking information are independent of their length of experience except for a slight and almost negligible positive correlation ($r = 0.15$) of the motive of maintaining professional competence (I) with experience. Other characteristics have correlated/associated significantly with many motives in the Table.[3]

Table 4.3 clearly shows that higher the status, stronger are the motives for doing own research (A) ($r_s = 0.94$), maintaining visibility among peers (B) ($r_s = 0.84$), publishing papers (K) ($r_s = 0.94$) and self-satisfaction (M) ($r_s = 0.75$) while seeking information.

From Table 4.4 it can be inferred that commensurating with higher qualifications, the space technologists are increasingly motivated to seek information for doing own research (A), achieving desired result in work (J), publishing papers (K) and continuing education (L) but decreasingly motivated by the prospects of getting promotion (E). The motives of having visibility among peers(B), maintaining an edge over competitors (C), gaining recognition (D) and self improvement (G) are also significantly related to the qualifications of the respondents, but not linearly.

It can be seen from Table 4.5 that the motives of doing own research (A), having visibility among peers(B), maintaining an edge over competitors (C), having knowledge (H), publishing papers (K) and self-

TABLE 4.2 : Association/Correlation Test Results of Motives of Seeking Information with (Selected) User-Characteristics

Motives	User-characteristics					
	Status	Qualifications	Nature of work	Specialisation	Experience	Professional Activities and Achievements Index
	(r)	*(x^2)*	*(x^2)*	*(x^2)*	*(r)*	*(r)*
	df = 8	*df* = 16	*df* = 24	*df* = 20	*df* = 325	*df* = 170
A. Own research	0.95*	52.94*	56.74*	28.58	0.04	0.27*
B. Visibility	0.82*	31.24*	40.42*	16.63	–0.01	0.21*
C. Edge over Competitors	0.42	27.69*	43.37*	16.79	–0.06	0.15*
D. Recognition	–0.01	31.46*	28.46	21.58	0.03	0.11
E. Promotion	–0.41	44.61*	29.40	31.87*	0.04	–0.27*
F. Project review	0.07	17.71	17.85	45.90*	0.02	0.01
G. Self-improvement	0.50	30.56*	30.50	18.29	0.01	0.02
H. For knowledge	0.40	22.74	40.55*	13.34	0.04	0.11
I. Professional competence	0.37	24.76	26.62	14.52	0.15*	0.16*
J. Desired result in work	0.84*	39.13*	35.06	18.42	–0.03	0.24*
K. Publishing Papers	0.94*	103.77*	46.21*	64.27*	0.04	0.39*
L. Continuing education	0.30	55.76*	28.67	39.98*	0.10	0.16*
M. Self-satisfaction	0.75*	24.89	40.53*	24.72	0.05	0.07

Key : * The association/correlation is statistically significant at 0.05 signifcance level (p) for a two tailed test.

Note ; The tabulated values of x^2 at 0.05 significance level (p) for a two tailed test for 16,24 and 20 degrees of freedom (df) are 26.296, 36.415 and 31.410 respectively.

TABLE 4.3 : Status Vs Motives of Seeking Information (Mean Score)

Motives	Status										
	A,B,C & D	E	F	G	H	I	J	K	L	M&N	Total	r_s
A. Own research	3.05	2.56	2.60	2.42	2.38	2.13	1.86	0.93	1.56	1.13	2.26	0.9515*
B. Visibility	2.21	1.93	1.77	1.77	1.86	1.50	1.78	1.43	1.75	1.38	1.77	0.8152*
C. Edge over Competitors	1.70	1.82	1.74	1.72	2.39	1.72	1.80	1.40	1.72	1.13	1.80	0.4182
D. Recognition	1.47	1.93	2.02	1.80	2.30	1.82	1.63	1.71	1.88	2.09	1.87	–0.0061
E. Promotion	1.20	2.11	2.18	2.07	2.33	2.05	2.62	2.71	2.71	2.18	2.20	–0.4121
F. Project Review	1.74	2.36	2.14	2.09	2.24	1.92	2.12	1.22	2.03	2.30	2.06	0.0667
G. Self improvement	3.26	2.83	3.25	3.21	3.28	3.27	3.22	2.96	2.70	2.69	3.14	0.5030
H. For knowledge	2.63	3.31	3.29	3.16	3.22	3.21	3.16	3.09	2.97	2.69	3.15	0.4030
I. Professional competence	3.05	3.36	3.09	2.96	2.90	3.03	2.93	3.17	3.06	2.69	3.03	0.3697
J. Desired result in work	3.47	3.30	3.36	3.08	3.10	3.28	3.15	2.57	3.00	2.80	3.15	0.8424*
K. Publishing Papers	2.13	1.55	1.75	1.47	1.49	1.17	0.83	0.30	0.94	0.25	1.31	0.9394*
L. Continuing education	2.37	1.44	1.91	1.77	2.36	2.31	2.46	1.05	2.00	0.91	1.96	0.2970
M. Self-satisfaction	3.33	3.09	3.13	2.82	2.86	3.09	3.09	2.88	2.78	1.90	2.97	0.7455*

Key : r_s Spearman rank order correlation; * the critical value of t at 0.05 significance level in a two tailed t-test for 8 degrees of freedom is 2.306 and correlations marked * are found valid at 0.05 significance level with t value exceeding 2.306.

Note : Status A to N are in the descending order of hierarchical rank.

TABLE 4.4 : Qualifications Vs Motives of Seeking Information (Mean Score)

Motives	Qualifications						
	U	D	B	M	P	Total	x^2
A. Own research	1.27	1.92	2.46	2.57	2.95	2.25	52.9394*
B. Visibiligy	1.27	1.62	1.95	1.83	1.77	1.76	31.2383*
C. Edge over competitors	1.32	1.81	1.97	1.86	1.32	1.79	27.6867*
D. Recognition	1.63	1.93	2.16	1.86	1.77	1.94	31.4623*
E. Promotion	2.58	2.38	2.24	1.99	1.32	2.19	44.6110*
F. Project review	1.86	2.03	2.17	2.07	1.76	2.05	17.7132
G. Self improvement	3.06	3.33	3.16	3.26	3.45	3.22	30.5637*
H. For knowledge	3.02	3.17	3.21	3.18	3.45	3.18	22.7385
I. Professional Competence	3.00	2.90	2.97	3.11	3.68	3.04	24.7600
J. Desired result in work	2.77	3.09	3.15	3.28	3.59	3.14	39.1348*
K. Publishing papers	0.46	1.09	1.40	1.62	2.50	1.32	103.7727*
L. Continuing education	1.50	2.47	2.03	1.85	2.36	1.99	55.7580*
M. Self-satisfaction	2.69	3.06	3.05	2.92	3.33	2.97	24.8875

Key : *, The association is statistically significant at $p = 0.05$ as the tabulated value of chi-square for $df = 16$ for two tailed test is 26.296; **Qualifications**-U, Undergraduate; D, Diploma holder; B,Graduate; M, Post-graduate; P, Doctorate.

Note : The contingency tables on which chi-square test is carried out are not shown.

satisfaction (M) are significantly related to the nature of work of the IST. But only doing own research as a motive is linearly and positively related to the nature of work of the space technologists in the ascending order of management/ supervision content or descending order of operational activity of their jobs.[4] The planners and system analysts are more likely to be motivated by maintaining visibility (B) and an edge over competitors (C), design and development personnel by an urge to publish (K), and managers and supervisors by self satisfaction (M) than others in seeking information.

Table 4.6 shows that the subject of specialisation[5] of the IST is signficantly related to the motives of getting promotion (E), facing

TABLE 4.5 : Nature of work Vs Motives of Seeking Information (Mean Score)

Motives	Nature of Work							Total	x^2
	A	B	C	P	D	E	F		
A. Own research	2.64	2.21	2.58	2.57	1.83	1.53	2.08	2.25	56.7429*
B. Visibility	1.93	2.06	1.75	2.26	1.33	1.80	1.85	1.74	40.4233*
C. Edge over competitors	1.91	2.10	1.71	2.21	1.47	1.93	1.71	1.77	43.3683*
D. Recognition	2.00	2.18	1.84	2.36	1.73	2.13	1.46	1.90	28.4586
E. Promotion	1.94	2.04	2.16	2.35	2.33	2.63	2.07	2.21	29.4029
F. Project review	2.06	2.19	2.14	2.25	1.98	1.72	1.68	2.06	17.8491
G. Self improvement	3.29	3.32	3.25	3.27	3.07	2.72	3.05	3.20	30.5014
H. For knowledge	3.26	3.09	3.25	3.34	3.08	3.17	3.05	3.17	40.5545*
I. Professional competence	1.89	3.11	3.10	3.09	2.84	2.80	2.94	3.03	26.6218
J. Desired result in work	3.22	3.19	3.30	3.43	2.94	2.84	3.11	3.15	35.0624
K. Publishing papers	1.45	1.40	1.56	1.33	0.88	1.13	1.20	1.29	46.2118*
L. Continuing education	2.03	2.01	2.00	1.74	1.96	2.28	1.86	1.99	28.6675
M. Self satisfaction	3.15	3.00	3.03	2.75	2.73	3.43	2.84	2.98	40.5318*

Key : * The association is statistically significant at $p = 0.05$ as the tabulated value of chi-square for $df = 24$ for two tailed test is 36.413; **Nature of work-** A, Management/Supervision; B, Planning/System Analysis; C, Design and Development; D, Fabrication and Testing; P, Both C and D; E; Operational Activity; F, Others.

Note : The contingency tables on which chi-square test is made are not shown.

TABLE 4.6 : Specialisation Vs Motives of Seeking Information (Mean Score)

Motives	Specialisation						Total	x^2
	A	C	E	F	H	I		
A. Own research	2.61	2.13	1.83	2.83	2.94	2.28	2.24	28.5822
B. Visibility	2.00	1.93	1.58	1.83	1.85	1.76	1.75	16.6335
C. Edge over competitors	1.76	2.07	1.57	1.83	2.00	1.80	1.76	16.7945
D. Recognition	1.88	2.13	1.86	2.28	2.07	1.84	1.89	21.5769
E. Promotion	1.69	2.33	2.29	1.79	3.07	2.22	2.20	31.8715*
F. Project review	1.74	2.12	1.97	1.94	3.20	2.10	2.07	45.8969*
G. Self improvement	3.18	3.33	3.20	3.58	3.00	3.18	3.20	18.2884
H. For knowledge	3.33	3.21	3.03	3.21	3.06	3.20	3.16	13.3403
I. Professional competence	2.95	2.22	1.57	3.28	3.18	2.94	2.52	14.5248
J. Desired result in work	3.15	3.39	2.98	3.33	3.20	3.17	3.13	18.4241
K. Publishing papers	1.60	1.41	0.79	2.22	2.08	1.35	1.29	64.2656*
L. Continuing education	1.46	2.12	1.60	2.72	2.62	2.05	1.94	39.9778*
M. Self-satisfaction	3.03	2.72	2.86	3.06	3.08	2.97	2.94	24.7205

Key : * The association is statistically significant as the tabulated value of chi-square at $p = 0.05$ for $df = 20$ in two tailed test is 31.410; **Specialisation**- A, Physics; C, Mathematics; E, Mechanical engineering; F, Aeronautical and Structural Engineering; H, Electrical Engineering; I, Electronics.

Note : The contingency tables on which chi-square test is made are not shown.

project review (F), publishing papers (K) and continuing education (L) in seeking information. Physicists within the population are least motivated by getting promotion (E), and facing project review (F). But the same are stronger motives of electrical engineers. In addition, electrical engineers as well as aeronautical and structural engineers have publishing papers (K) and continuing education (L) as stronger motives than others.

Table 4.2 also reveals that the more professionally active a space technologist is, the more likely he is to be motivated, while seeking informaion, to do research independently (A) (r = 0.27), maintain visibility among peers (B) (r = 0.21), have an edge over competitors (C) (r = 0.15), build professional competence (I) (r = 0.16), achieve desired result in work (J) (r = 0.24), publish papers (K) (r = 0.39) and continue education (L) (r = 0.16), but he is less likely to be motivated by getting promotion (E) (r = -0.27).

4.3 Purposes of Seeking Information

What the Indian space technologists felt about their purposes of seeking information is reflected in the responses elicited through questionnaire (Q.No.2) and tabulated in Table 4.7. As could be seen from the Table, the purposes for which a considerable number of respondents do not seek information are : writing papers and delivering lectures (G-34.3 per cent), data treatment and processing (D-27.4 per cent), participating in seminars and conferences (H-26.0), crystallising broad and vague assertions (M-24.2 per cent), designing and development of products and components (B-21.8 per cent) and programme/activity planning (A-20.9 per cent). More than 80 per cent of the respondents (as per the Table 4.7) seek information for all the remaining purposes noted in the Table 4.7.

A large number of the respondents attributed the highest weightage of 'highly used' (code 4) for purposes of 'keeping abreast with latest developments in the field' (K-130), 'designing and development of products and components' (B-99) and 'broadening area of attention and reviewing work done in related areas' (J-71). It may also be observed from the Table that while as many as 113 felt that they do not seek information for 'Designing and development of products and components', a considerable number of respondents (*i.e.* 99) consider that to be the strongest purpose of seeking information, indicating, thereby, that this purpose might be strongly related to nature of work and other attributes of the respondents. Figure 4.2 depicts the responses on the five point

TABLE 4.7 : Purposes of Seeking Information

Purposes	Not Applicable (0)		Applicable				Total		Mean	S.D.
	No.	%	1	2	3	4	No.	%		
A. Program/activity planning	107	20.9	121	113	107	64	405	79.1	1.80	1.33
B. Designing and development of products & components	113	21.8	71	102	133	99	405	78.2	2.07	1.42
C. Setting up and use of equipment	84	20.1	83	121	149	65	418	79.9	2.06	1.28
D. Data treatment and processing	139	27.4	111	109	112	37	369	72.6	1.60	1.29
E. Checking authenticity of available results/data/ information	92	18.1	117	166	103	30	416	81.9	1.73	1.15
F. Checking and evaluating ones/own results	59	11.5	100	130	168	55	453	88.5	2.31	0.73
G. Writing paper and delivering lecture	176	34.3	135	100	80	22	337	65.7	1.29	1.21
H. Participating in seminars and conferences	134	26.0	146	137	81	17	381	74.0	1.42	1.13
I Guiding team and sharing information with members of the team	70	13.6	105	132	164	43	444	86.4	2.01	1.18

(Contd.)

TABLE 4.7 (Contd.) : Purposes of Seeking Information

Purposes	Not Applicable(0)		Applicable				Total		Mean	S.D.
	No.	%	1	2	3	4	No.	%		
J. Broadening area of attention and reviewing work done in related areas	31	6.1	103	149	157	71	480	93.9	2.26	1.12
K. Keeping abreast with latest developments in the field	17	3.2	73	73	231	130	507	96.8	2.73	1.09
L. Orienting one's/own work with the existing body of knowledge	29	5.8	95	119	197	62	473	94.2	2.33	1.10
M. Crystallizing broad and vague assertions	117	24.2	114	130	106	16	366	75.8	1.57	1.15
N. Evolving innovative ideas/ techniques	55	11.1	123	140	135	43	441	88.9	1.98	1.14
O. Others	181	91.4	5	4	7	1	17	8.6	0.19	0.69

Key : *, O, Not used; 1, Rarely used. 2, Occasionally used; 3. Frequently used; 4, Highly used.

Notes : 1. Three hundred and twenty eight respondents have left open-ended category 'other's (O) blank.

2. Not answered and invalid responses which varied from 11 to 52 are excluded.

3. The mean score for anonymous responses worked out to 1.87,2.17, 1.80, 1.24, 1.31, 2.10, 0.69, 0.93, 1.50, 2.17, 2.57, 2.07, 2.10, 1.79, 0.17 respectively to A to O.

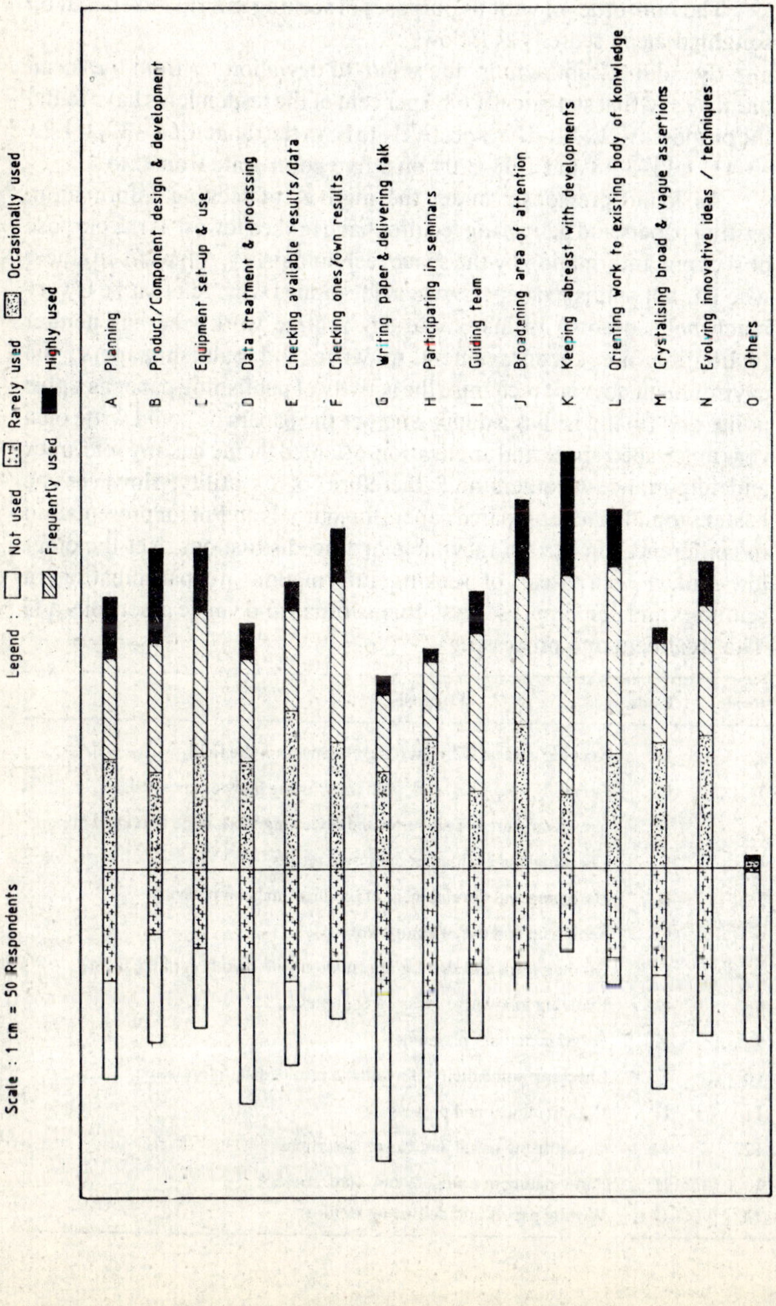

Fig 4.2 : Purposes of Seeking Information

scale to various purposes and emphasises the above results.

The rank order of various purposes of seeking information based on weighted mean score is as follows :

By adding/subtracting one standard deviation to/ from the mean one may say that statistically 68.3 per cent of the respondents have found the purposes K, L, J and F respectively to be in the range 1.64 – 3.82, 1.23 –3.43, 1.14–3.38 and 1.58–3.04 on a five point scale from 0 to 4.

As found previously under the motives of seeking information, writing papers and delivering lectures have ranked lowest as the purpose of seeking information by the space technologists.[8] A typical engineer who has not published papers, when questioned why, felt that he is very much bogged down in his day-to-day routine work ; he has neither inclination nor encouragement to write and publish papers; the environment does not recognise the activity of publishing paper as a plus point; and finally he has a doubt whether the papers he could write on a very much specialised and application-oriented theme has any relevance and importance to others and, therefore, of its utility. However, he hastens to add that he prepared papers for journals and/or for presentation in conferences in certain inevitable or forced situations. Yet the other low-ranked purposes of seeking information are participating in seminars and conferences, crystallising broad and vague assertions and data treatment and processing[9, 10]

Rank	Code	Purposes
1	K	Keeping abreast with latest developments in the field_
2	L	Orienting ones/own work with the exisitng body of knowledge
3	J	Broadening area of attention and reviewing work done in related areas
4	F	Checking and evaluating ones/own results
5	B	Designing and development of products and components[6]
6	C	Setting up and use of equipment
7	I	Guiding team and sharing information with members of the team
8	N	Evolving innovative ideas / techniques
9	A	Program/activity planning.
10	E	Checking authenticity of available results/data/information
11	D	Data treatment and processing
12	M	Crystallising broad and vague assertions
13	H	Participating in seminars and conferences [7]
14	G	Writing papers and delivering lectures

The foremost purpose of seeking information by the space technologists, (*viz.*, keeping upto date in the field) is in conformity with what Menzel (Columbia University, 1958, p. 67 and 80) (Keeping abreast of current developments) and Voigt (1959, p. 177-179) (Current approach) have found earlier. However in the study of Slater and Fisher (1969, p. 32) the purpose of 'keeping abreast' occupied second rank whereas the purpose of knowing past work in the field (existing body of knowledge) has ranked first. The interchange of these first and second ranked purposes of seeking information in the present study could be attributed to the differences in methodology and population characteristics. The purpose of broadening the area of attention and reviewing the work done in related areas has taken the next priority among the purposes of seeking information. A respondent remarks that "information is gathered to increase the general sense of awareness about the state-of-the-art in the area of work to see how it can be applied in the work and to identify and pursue a desirable degree of professional competence".

4.4 Correlation of Purposes of Seeking Information with (Selected) User-characteristics

Table 4.8 presents the results of association/correlation tests of purposes of seeking information with six user-characteristics, namely, status, qualifications, nature of work, specialisation, length of experience and professional activities and achievements. Tables 4.9, 4.10, 4.11 and 4.12 display the mean score of responses against the status, qualifications, nature of work and specialisation respectively. As found in the case of motives of seeking information, the purposes of seeking information are found to be independent of the length of experience except for a slight and almost negligible negative correlation ($r = -0.16$) with the purpose of data treatment and processing (D) and a low positive correlation ($r = 0.22$) with the purpose of guiding team (I). However, unlike motives of seeking information, all the purposes of seeking information (except design and development of products /components) have exhibited statisticallysignificant relationship with professional activities and achievements of the IST.

Table 4.9 indicates that status has a very high correlation with all the purposes of seeking information. As the status of the space technologists increases, decreasingly less information is sought for the purpose of equipment setup and use (C) ($r_s = -0.74$) and increasingly more information is sought for all other purposes.

TABLE 4.8 : Association/Correlation Test Results of Purposes of Seeking Information with Selected User-Characteristics

	Purposes	Status	Qualifications	Nature of work	Specialisation	Experience	Professional Activities and Achievements Index
		(r_s) $df = 8$	(x^2) $df = 16$	(x^2) $df = 24$	(x^2) $df = 20$	(r) $df = 384$	(r) $df = 191$
A.	Planning	0.94*	108.49*	82.81*	61.22*	0.05	0.42*
B.	Product/component design and development	0.85*	59.87*	91.29*	50.42*	0.06	0.01
C.	Equipment setup and use	–0.74*	42.00*	90.80*	58.20*	0.04	-0.16*
D.	Data treatment and processing	0.87*	85.81*	50.44*	94.25*	–0.16	0.24*
E.	Checking available results/data	0.95*	79.14*	42.15*	38.33*	–0.07	0.19*
F.	Checking ones's/own results	0.85*	69.25*	37.78*	32.14*	–0.05	0.22*
G.	Writing paper and delivering talk	0.99*	139.29*	85.81*	57.70*	0.09	0.46*
H.	Participating in seminars	0.93*	104.31*	64.41*	60.21*	0.10	0.39*
I.	Guiding team	0.95*	93.16*	72.49*	30.77	0.22*	0.36*
J.	Broadening area of attention	0.98*	120.37*	93.68*	45.42*	0.04	0.37*
K.	Keeping abreast with developments	0.98*	138.26*	78.86*	63.97*	0.05	0.35*
L.	Orienting work to existing body of knowledge	0.93*	87.83*	81.45*	32.01*	0.09	0.33*
M.	Crystallising broad and vauge assertions	0.94*	65.41*	48.47*	51.62*	0.10	0.42*
N.	Evolving innovative ideas/techniques	0.98*	77.60*	70.14*	36.92*	0.09	0.28*

Key : *, The association/correlation is statistically significant at 0.05 significance level(p) for a two tailed test.

Note : The tabulated values of chi-square at 0.05 significance level (p) for a two tailed test for 16,24 and 20 degrees of freedom *(df)* are 26.296, 36.415 and 31.410 respectively.

TABLE 4.9 : Status Vs Purposes of Seeking Information (Mean Score)

Purposes	Status										
	A,B,C and D	E	F	G	H	I	J	K	L	M&N	Total	r_s
A. Planning	3.20	2.62	2.31	1.78	1.76	1.45	1.48	0.61	1.03	1.24	1.68	0.9394*
B. Product-component design and development	2.70	2.37	2.58	2.31	2.25	1.74	1.77	1.52	1.63	1.94	2.11	0.8545*
C. Equipment setup and use	1.75	1.94	2.14	1.93	1.94	1.94	2.14	1.96	2.55	2.78	2.06	–0.7424*
D. Data treatment and processing	1.80	1.65	1.89	1.78	1.77	1.58	1.43	1.05	0.82	0.57	1.57	0.8727*
E. Checking available results/data	2.40	1.86	2.11	1.99	1.68	1.67	1.48	0.73	1.21	0.75	1.71	0.9515*
F. Checking ones's/own results	2.65	2.12	2.52	2.37	2.19	2.29	1.77	1.38	1.70	1.05	2.01	0.8545*
G. Writing paper and delivering talk	2.45	2.00	1.86	1.38	1.27	0.69	0.98	0.65	0.39	0.17	1.24	0.9879*
H. Participating in seminars	2.15	2.12	1.75	1.48	1.60	0.85	1.15	0.61	0.95	0.50	1.38	0.9273*
I. Guiding team	2.95	3.02	2.42	2.17	1.92	1.56	1.44	1.35	1.50	0.89	1.99	0.9515*
J. Broadening area of attention	3.00	2.96	2.63	2.51	2.22	2.09	1.74	1.55	1.76	1.38	2.27	0.9758*
K. Keeping abreast with developments	3.40	3.40	3.22	3.06	2.75	2.48	2.16	1.83	2.10	1.22	2.92	0.9848*
L. Orienting work to existing body of knowledge	2.75	2.92	3.26	2.56	2.23	2.23	1.94	1.59	1.92	1.00	2.40	0.9303*
M. Crystallising broad and vauge assertions	2.40	2.18	1.92	1.51	1.80	1.20	1.28	0.81	0.86	0.82	1.53	0.9394*
N. Evolving innovative ideas/ techniques	2.55	2.53	2.45	1.97	2.27	1.66	1.55	1.33	1.36	1.13	1.97	0.9758*

Key : r_s Spearman rank order correlation; *, All correlations are found to be valid at 0.05 significance level as the critical value of t at 0.05 significance level in a two tailed t-test for 8 degrees of freedom is 2.306 and all of them exceed this critical value.

Note : Status A to N are in the descending order of hierarchical rank.

Data in Table 4.10 shows that the educational level of the IST has a linear positive relation with seeking information for all the purposes except equipment setup and use (C) and product/component design and development (B), where the relations are significat but not linear.

As could be seen from Table 4.11, with the increase in the management/supervision content of the job, space technologists have increasingly sought information for the purpose of participating in seminars, symposia and conferences (H), guiding team (I), keeping abreast with latest developments (K), orienting one's work to existing body of knowledge (L), crystallising broad and vague assertions (M) and evolving innovative ideas/techniques (N). The design and development personnel have sought information more for the purpose of product/ component design and development (B), data treatment and processing (D) and checking one's results (E) than other purposes. Personnel involved in operational activity, fabrication and testing have sought information more for equipment setup and use than others.[11]

Lastly, Table 4.12 shows that specialisation of the IST is significantly related to all the purposes of seeking information except guiding team (I). Physicists within the population have sought information more for the purposes of planning the work (A), writing papers and delivering talks (G), participating in seminars (H) and keeping abreast with latest developments (K) than other specialists. Mathematicians have sought information more for the purposes of planning (A), data treatment and processing (D), broadening area of attention (J), crystallising broad and vague assertions (M) than for other reasons. Mechanical engineers have sought information more for the purpose of product/component design and development (B) than for other purposes. Aeronautical and structural engineers have sought information for checking results (E&F), writing papers and delivering talks (G), participating in seminars (H), guiding team members (I), broadening area of attention (J), keeping abreast with developments (K), orienting to existing body of knowledge (L), crystallising broad and vague assertions (M) and evolving innovative ideas (N) then other specialists. The equipment setup and use (C) as well as evolving innovative ideas/techniques (N) are the strong purposes of electrical engineers in seeking information compared to other specialists. Like mechanical engineers, electronics engineers also sought more information for the purpose of product/component design and development (B) than other purposes.

Table 4.8 indicates that as the experience of the IST increases, they seek information increasingly more for the purpose of guiding team

TABLE 4.10 : Qualification Vs Purposes of Seeking Information (Means Score)

Purposes	Qualifications					Total	x^2
	U	D	B	M	P		
A. Planning	0.78	1.38	0.56	2.15	3.00	1.71	108.4886*
B. Product/component design and development	1.72	2.05	2.56	2.16	2.17	2.18	59.8663*
C. Equipment setup and use	2.29	1.74	2.09	1.73	2.08	2.08	42.0011*
D. Data treatment and processing	0.63	1.33	1.53	1.90	2.21	1.48	85.8086*
E. Checking available results/data	0.91	1.51	1.84	2.04	2.30	1.69	79.1436*
F. Checking ones's/own results	1.31	1.92	2.24	2.33	2.60	2.06	69.2453*
G. Writing paper and delivering talk	0.35	0.70	1.26	1.59	2.56	1.16	139.2893*
H. Participating in seminars	0.58	1.03	1.48	1.55	2.30	1.30	104.3144*
I. Guiding team	1.27	1.54	2.03	2.32	3.17	1.93	93.1639*
J. Broadening area of attention	1.44	2.00	2.47	2.62	3.13	2.72	120.3703*
K. Keeping abreast with developments	1.76	2.46	2.96	3.08	3.43	2.70	138.2616*
L. Orienting work to existing body of knowledge	1.45	2.15	2.40	2.57	3.04	2.26	87.8268*
M. Crystallising broad and vauge assertions	0.78	1.30	1.61	1.65	2.50	1.46	65.4142*
N. Evolving innovative ideas/ techniques	1.21	1.71	2.06	2.13	2.91	1.91	77.6042*

Key : *, The association is statistically significant at $p = 0.05$ as the tabulated value of chi-square at $p = 0.05$ for 16 degree of freedom is 26.296; *Qualifications*— U, Undergraduates; D, Diploma holder; B, Graduate; M, Postgraduate; P, Dectorate.

Note : The contingency tables on which chi-square test is made are not shown.

TABLE 4.11 : Nature of Work Vs Purposes of Seeking Information (Mean Score)

Purposes	Nature of Work								
	A	B	C	P	D	E	F	Total	x^2
A. Planning	2.78	2.30	2.58	1.73	1.27	1.59	1.16	1.74	82.8130*
B. Product/component design and development	2.05	1.70	2.75	2.80	2.00	2.00	1.32	2.23	91.2928*
C. Equipment setup and use	2.17	1.12	2.10	2.48	2.51	2.72	1.42	2.10	90.7967*
D. Data treatment and processing	1.71	1.60	1.74	1.55	1.02	1.40	1.40	1.50	50.4436*
E. Checking available results/data	1.95	2.02	1.70	1.90	1.30	1.30	1.84	1.70	42.1477*
F. Checking ones's/own results	2.10	2.40	2.30	2.10	1.70	2.00	2.20	2.10	37.7759*
G. Writing paper and delivering talk	1.78	1.64	1.29	1.14	0.72	0.76	0.79	1.15	85.8078*
H. Participating in seminars	1.83	1.71	1.35	1.04	0.98	0.91	0.95	1.26	64.4101*
I. Guiding team	2.68	2.46	1.94	2.08	1.50	1.32	2.00	1.91	72.4949*
J. Broadening area of attention	2.83	2.84	2.42	2.56	1.70	1.90	2.21	2.70	93.6799*
K. Keeping abreast with developments	3.17	3.01	2.91	3.10	2.09	2.64	2.70	2.70	78.8646*
L. Orienting work to existing body of knowledge	2.88	2.60	2.44	2.56	1.68	2.03	2.35	2.26	81.4488*
M. Crystallising broad and vauge assertions	1.90	1.70	1.55	1.50	1.38	0.93	1.37	1.43	48.4652*
N. Evolving innovative ideas techniques	2.29	2.28	2.08	2.12	1.44	1.43	1.80	1.89	70.1418*

Key : *, The association is statistically significant at $p = 0.05$ as the tabulated value of chi-square at $p = 0.05$ for 24 degree of freedom is 36.415; *Nature of Work*-A, Management/Supervision; B, Planning/System Analysis; C, Design and Development; D, Fabrication and Testing; P, Both C and D; E, Operational Activity; F, Others.

Note : The contingency tables on which chi-square test is made are not shown.

TABLE 4.11 : Specialisation Vs Purposes of Seeking Information (Mean Score)

Purposes	Specialisation						Total	x^2
	A	C	E	F	H	I		
A. Planning	2.38	2.30	1.23	2.16	2.06	1.77	1.72	61.2203*
B. Product/component design and development	2.12	1.11	2.28	2.11	1.69	2.48	2.28	50.4204*
C. Equipment setup and use	1.95	1.00	2.13	1.00	2.44	2.30	2.11	58.2008*
D. Data treatment and processing	1.83	3.17	0.96	1.68	1.38	1.49	1.44	94.2485*
E. Checking available results/data	2.02	2.53	1.44	2.05	1.20	1.65	1.66	38.3269*
F. Checking ones's/own results	2.17	2.89	1.77	2.63	1.87	2.04	2.03	32.1396*
G. Writing paper and delivering talk	1.76	1.53	0.66	1.74	1.06	1.19	1.12	57.6997*
H. Participating in seminars	1.86	1.60	0.90	1.84	1.00	1.36	1.29	60.2115*
I. Guiding team	2.20	2.11	1.68	2.58	2.00	1.96	1.93	30.7663*
J. Broadening area of attention	2.56	2.47	1.97	3.00	2.31	2.30	2.25	45.4206*
K. Keeping abreast with developments	2.98	2.85	2.23	3.42	2.63	2.82	2.68	63.9676*
L. Orienting work to existing body of knowledge	2.49	2.59	2.01	2.63	2.31	2.25	2.23	32.0134*
M. Crystallising broad and vauge assertions	1.85	1.88	1.02	2.11	1.50	1.53	1.44	51.6247*
N. Evolving innovative ideas/ techniques	2.00	2.00	1.65	2.26	2.31	1.95	1.89	36.9229*

Key : *, The association is statistically significant at p = 0.05 as the tabulated value of chi-square for df = 20 at p = 0.05 in a two tailed test is 31.410; *Specialisation*-A, Physics; C, Mathematics; E, Mechanical Engineering; F, Aeronautical and Structural Engineering; H, Electrical Engineering; I, Electronics.

Note : The contingency tables on which chi-square test is made are not shown.

mater (I) (r = 0.22) but decreasingly less for the purpose of data treatment and processing (D) (r = –0.16). As professional activities and achievements of the IST increased, information is sought less for the purpose of equipment setup and use (C) (r = –0.16) and more for all other purposes except for product/compoent design and development.

NOTES

1. This motive is naturally found to be significantly associated with the number of part time studies undertaken by the respondents ($x^2 = 15.51$, $df = 9$, $p < 0.05$).
2. Out of 28 who have responded to 'other motivational factors' (N), majority have either not specified or vaguely specified such motives. Some interesting responses noticed are-'I enjoy a good book', 'interest in new development in the field', 'understanding self and world around', 'to help others', 'encouragement from seniors', 'to identify opportunity and new areas of activity needed for organisation' and 'to spend time or to be more correct to kill the time'. In addition, during the interview certain other motives like 'curiosity of knowing things', 'urge to learn', 'maintaining reputation', 'personal interest', 'to be on the state-of-the-art' and 'to improve the present method' are noticed.
3. All the motives which ranked high (with high mean score) in Table 4.1 (*i.e.*, G,H,I,J) are not significantly related to many user-characteristics but such motives which scored low mean (*e.g.* K) are significantly related to more user- characteristics.
4. The nature of works of the space technologists in the ascending order of management/ supervision content or descending order of operational activity of their jobs are : E — Operational activity, D — Fabrication and testing, C — Design and development, B — Planning/System analysis, A — Management/Supervision.
5. All specialisations other than those given in Table 4.6 where the data are insufficient for an effective chi-square test are ignored for correlationalstudy throghout the book. The specialisations mechanical engineering, electrical engineering and electronics have disproportionately more technicians of lower status, qualifications, management/ supervision content in the nature of work, experience and professional activities and achievements. Hence they are likely to contribute disproportionately more to the mean score of these three specialisations.
6. The purposes of designing and development of products/ components and setting up and use of equipment are positively and moderately interrelated (r = 0.4).
7. A moderate positive correlation (r = 0.44) is found between the purpose of seeking information for participating in seminars and conferences and the number of seminars, symposia and conferences attended during the last year.
8. 'To write and publish' as a motive is correlated moderately and positively (r = 0.53) with the purpose of 'writing papers and delivering talks' in seeking information. The purpose 'writing papers and delivering talks' is also correlated moderately and positively with the number of papers published so far (r = 0.44) and the number of lectures delivered during the last year (r = 0.47).
9. The 17 responses noted under 'other purposes' (0) in Table 4.7 are quite varied like 'comparing and consolidating own experience with others', 'preparing documents and reports', 'cross checking with overall system design', 'pursuing formal education', 'for general knowledge', 'time spending', 'clarifying colleagues questions', 'imple-

menting the latest techniques and making systems highly flexible', 'taking preventive measures', 'routine maintenance', and 'conducting meetings with colleagues to improve present status'.

10. The maximum unanswered responses are found for 'crystallising broad and vague assertions' (M) (52 respondents), 'evolving innovative ideas/techniques' (N) (39 respondents) and 'orienting one's/own work with the existing body of knowledge' (L) (33 respondents). The invalid responses did not exceed one for any of the purposes in Table 4.7.

11. Interestingly crystallising broad and vague assertions (M) and participating in seminars (H) as purposes of seeking information have occupied last positions in the overall ranking. These are the purposes for which information is sought mainly by managers/supervisors and planners/system analysts.

5

INFORMATION REQUIREMENTS OF THE INDIAN SPACE TECHNOLOGISTS

(Nature and Type of Information Sought by the Indian Space Technologists)

5.1 Nature and Type of Information Required

The felt and expressed information requirements of the Indian space technologists, in respect of nature and type of information, have been ascertained through the questionnaire (Q.No.1) and presented in Table 5.1. It is clear from the Table that the respondents not seeking 'Statistical, economic, business and general information' (J-26.3%), 'Standard and patent specifications and codes of practice' (G-24.1%), 'Computer program and model building information' (F-23.0%) and 'State-of-the-art and review literature' (A-19.4%), are quite considerable. Other types of information as shown in Table 5.1 are required by more than 90 per cent of the respondents.

On the other hand more respondents have highly sought 'Scientific and technical news' (I-147), 'Experimental designs, results and applications' (C-102), 'State-of-the-art and review literature' (A-84) and 'theoretical background/basic scientific and technical information' (B-84), than other types of information. An interesting feature is that, the state-of-the-art and review literature is not required by 101 respondents while it is highly sought after bv 84. The histograms in Fig. 5.1, by showing 'Not required' and 'Rarely required' on one side and the remaining on the other (positive) side, pictorially depict the responses.[1]

The rank order of the nature and type of information required by the

TABLE 5.1 : Nature and Type of Information Required

Nature and Type of Information	Not Required (O)		Required						Mean	S.D.
							Total			
	No.	%	1	2	3	4	No.	%		
A. State-of-the-art and review literature	101	19.4	94	97	144	84	419	80.6	2.03	1.37
B. Theoretical background/basic scientific and technical information	7	1.3	35	126	280	84	525	98.7	2.75	0.85
C. Experimental designs, results and applications	29	5.5	59	129	207	102	497	94.5	2.56	1.09
D. Methods, processes and procedures	46	8.8	69	170	173	65	477	91.2	2.27	1.12
E. Product, material, equipment and apparatus information	43	8.2	79	195	137	69	480	91.8	2.21	1.11
F. Computer programs and model building information	120	23.0	147	135	75	45	402	77.0	1.57	1.24
G. Standard and patent specifications and codes of practice	125	24.1	163	121	77	32	393	75.9	1.47	1.19
H. Physical, technical and design data	16	3.1	70	178	198	59	505	96.9	2.41	0.96
I. Scientific and technical news	10	1.9	34	147	192	147	520	98.1	2.82	0.96
J. Statistical, economic, business and general information	136	26.3	251	122	36	9	382	73.7	1.16	1.20
K. Others	17	36.2	10	4	9	7	30	63.8	1.55	1.50

Key : 0, Not required; 1, Rarely required; 2, Occasionally required; 3, Frequently required; 4 Highly sought-after; S.D. Standard Deviation.

Note : 1. Three hundred and fifteen have left the open-ended category 'Others' (K) blank.

2. Invalid and not answered responses upto a maximum of 17 in any category are excluded.

3. The mean score for anonymous responses worked out to 1.55, 3.10, 2.30, 2.07, 2.03, 1.14, 1.43, 2.27, 2.57, 0.86 and 0.05 respectively for categories A to K.

Rank	Code	Nature and Type of Information
1	I	Scientific and technical news
2	B	Theoretical background/basic scientific and technical information
3	C	Experimental designs, results and applications
4	H	Physical, technical and design data
5	D	Methods, processes and procedures[2]
6	E	Product, material, equipment and apparatus information
7	A	State-of-the-art and review literature
8	F	Computer programs and model building information
9	G	Standard and patent specifications and codes of practice
10	J	Statistical, economic, business and general information

space technologists, based on weighted mean, is given below. The top ranks occupied by S & T news (I) and basic S & T information (B) are in congruence with the top ranking motives and purposes like acquiring knowledge about latest developments, self-improvement and keeping update in the field.

Adding/subtracting one standard deviation to/from mean in Table 5.1, one can statistically say that, nearly 68 per cent of respondents have sought 'Scientific and technical news' (I) with the weightage ranging from 1.86 to 3.78, 'Experimental designs, results and applications'(C) with weightage ranging from 1.47 to 3.65; 'physical, technical and design data' (H) with weightage from 1.45 to 3.37; 'methods, processes and procedures information' (D) with weightage from 1.15 to 3.39; and 'product, materials, equipment and apparatus information'(E) with weightage ranging from 1.1. to 3.32.

The top ranked need for basic S & T information and S&T news by the space technologists is close to that found by Shuchman (1981, p. 32-33; 1982, p.106–109) about American industrial engineers, Wood (1967, p. 212) about British mechanical engineers, Slater and Fisher (1969, p. 32-33, 36, 47 and 49) about users of British technical libraries and Rawdin (1975, p. 41–42) about users of American technical libraries.

The next in the rank order, the space technologists sought experimental designs, facts, data, methods, procedures and product information which again is close to the second ranked type of information in the studies of Shuchman, (1981, 1982), Rawdin (1975) and Stater and Fisher (1969).

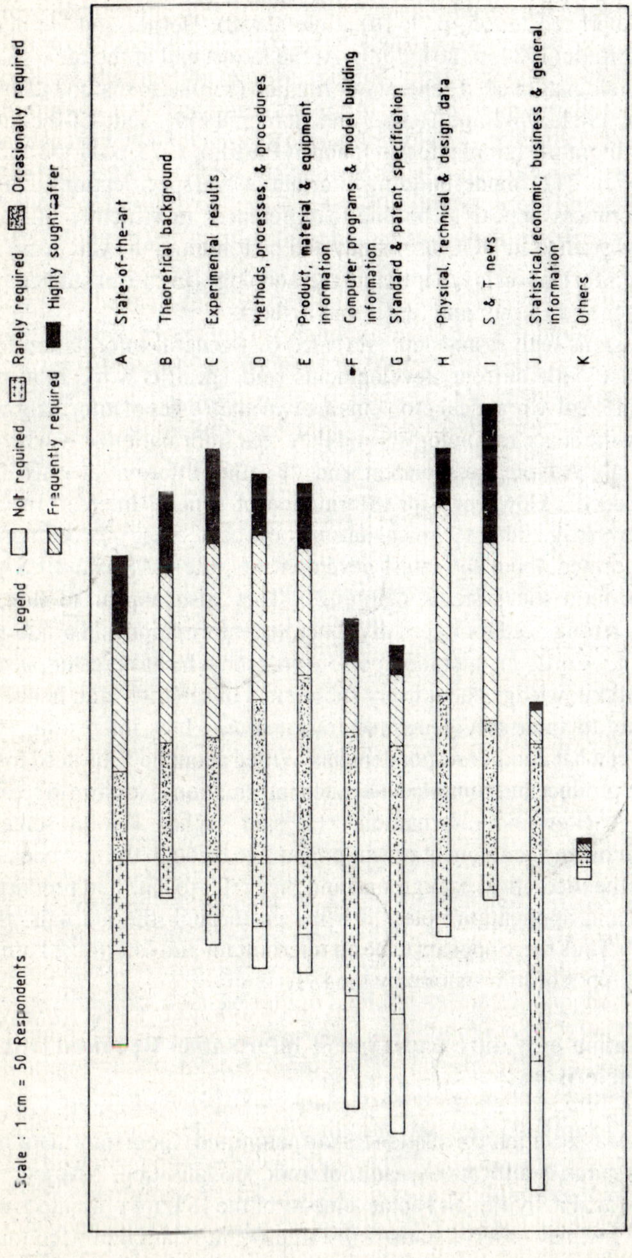

Fig. 5.1 : Nature and Type of Information Required

However, this type of information ranked topmost in the studies of DOD users by Auerbach (1965, p. 1–19), Cole (1958), Herner and Herner (1959) and Raitt (1984, p. 204–208). At the lowest end of the rank, the space technologists needed general information (similar to that found by Shuchman, 1981, 1982 and Herner and Herner, 1959), state-of-the-art and review literature (similar to that found by Rawdin, 1975, p. 41–42 and Wood, 1967, p. 212), model building information and specifications. The above differences appear to be due to differences in structure of the populations studied in DOD user study and by Raitt and the way nature and type of information is grouped and methodology (in case of studies of Cole, Herner and Herner and Slater and Fisher).

Discussions with respondents revealed that general information to keep abreast with current developments and specific work related information to solve practical problems are two main types of information required by the space technologists and they seek information only when it is essential. A typical respondent said "I gather information only if utmost needed". However, for information of general interest, they "browse news magazines, technical journals and conference proceedings to keep informed about the latest developments, the NASA and ESA reports for futuristic systems planning. They also appear to have moderately strong need for internally generated information and product information. A mission operation person writes that he has greater need to "have full knowledge about every sub-system of satellite" and hence turns mostly to internally generated information. It is interesting to discern from what another respondent has written about how his need for equipment, product and component-related information leads him to seek theoretical background information. He said "When I want some information on any component or equipment I go through the text books to build up the theoretical background and then refer to data and product catalogues and application notes. Then I go through the journals, if necessary." Thus there appears to be an intermittant and intermixed use of different types of information by the IST.

5.2 Correlation of Nature and Type of Information Required with User-characteristics

The results of association/correlation tests of nature and type of information sought with status, qualifications, nature of work, specialisation, experience and professional activities and achievements of the IST are presented in Table 5.2. One may observe statistically significant relation of need for state-of-the-art literature (A), computer programs and model building information (F) and standard and patent specifications (G) with all the above characteristics. The requirements of physical, technical and design data (H) are found to be independent of all the above user characteristics, except status. Tables 5.3, 5.4, 5.5 and 5.6 respectively present the weighted mean score of nature and type of information sought versus status, qualifications, nature of work and specialisation.

Table 5.3 reveals that as the status of a user raises, his requirement for state-of-the-art and review literature (A) (r_s = 0.95), experimental results (C) (r_s = 0.92), product, material and equipment information (E)

TABLE 5.2 : Association/Correlation Tests Result of Nature and Type of Information Required with (Selected) User-characteristics

Nature and Type of Information	Status (r_s) $df = 8$	Qualifications (x^2) $df = 16$	Nature of work (x^2) $df = 24$	Specialisation (x^2) $df = 20$	Experience (r) $df \sim 390$	Professional Activities & Achievements Index (r) $df \sim 192$
A. State-of-the-art	0.95*	176.56*	97.86*	83.96*	0.16*	0.42*
B. Theoretical background	0.55	56.20*	29.50	20.11	–0.12*	0.02
C. Experimental results	0.92*	44.35*	27.27	39.51*	0.02	0.09
D. Methods, processes and procedures	0.58	51.20*	45.56*	29.79	-0.03	0.13*
E. Product, material and equipment informatation	0.98*	27.25*	22.26	43.01*	0.08	0.08
F Computer programs and model building information	0.72*	83.23*	46.22*	104.30*	–0.21*	0.23*
G. Standard and patent specifications	–0.72*	73.46*	39.36*	67.16*	0.12*	–0.21*
H. Physical, technical and design data	0.72*	13.25	35.79	29.48	0.04	–0.03
I. S & T news	0.43	22.80	21.94	39.98*	0.13*	0.07
J Statistical economic, business and general information	0.71*	38.30*	41.48*	32.20*	0.05	0.05

Key : *, The association/correlation is statistically significant at 0.05 significance level (p) for a two-tailed test.

Note : The tabulated values of x^2 at 0.05 significance level (p) for a two-tailed test for 16, 24 and 20 degree of freedom *(df)* are 26.296, 36.415 and 31.410 respectively.

TABLE 5.3 : Nature and Type of Information Required Vs Status (Mean Score)

Nature and Type of Information	Status										
	A,B,C & D	E	F	G	H	I	J	K	L	M&N	Total	r_s
A State-of-the-art	3.29	3.37	3.06	2.27	1.92	1.44	1.00	0.88	1.21	0.41	2.00	0.9515*
B Theoretical background	2.81	2.75	2.96	2.74	2.80	2.81	2.82	2.31	2.63	2.11	2.74	0.5515
C. Experimental results	2.76	2.85	2.87	2.60	2.67	2.39	2.44	1.84	2.23	1.82	2.53	0.9152*
D. Methods, processes and procedures	2.05	2.51	2.37	2.25	2.33	2.25	2.32	1.92	2.10	1.60	2.25	0.5758
E. Product, material and equipment information	2.62	2.27	2.54	2.21	2.19	2.04	2.19	2.04	1.93	1.79	2.16	0.9758*
F. Computer programs and model building information	1.52	1.73	1.84	1.65	1.56	1.53	1.55	1.04	0.95	0.47	1.49	0.7212*
G. Standard and patent specifications	0.95	1.38	1.30	1.08	2.14	1.65	1.83	2.04	2.00	2.00	1.52	-0.7242*
H. Physical, technical and design data	2.38	2.46	2.54	2.45	2.39	2.38	2.43	2.21	2.30	2.44	2.36	0.7227*
I S & T news	2.95	3.02	2.90	2.61	2.67	2.76	2.82	2.52	2.88	2.89	2.72	0.4303
J. Statistical, economic, business and general information	1.29	1.38	1.30	1.04	1.27	1.02	1.27	1.09	1.15	0.65	1.14	0.7121*

Key : r_s Spearman rank order correlation; *, the critical value of t at 0.05 significance level in a two-tailed t-test for degree of freedom is 2.306 and the correlations marked * are found valid at 0.05 significance level as their t-value exceed 2.306

Note : Status A to N are in the descending order of hierarchical rank.

TABLE 5.4 : Nature and Type of Information Required Vs Qualifications (Mean Score)

Nature and Type of Information	Qualifications					Total	x^2
	U	D	B	M	P		
A. State-of-the-art	0.93	1.21	2.18	2.58	3.52	1.96	176.5601*
B. Thoretical back-ground	2.31	2.88	2.86	2.75	3.35	2.76	56.2022*
C. Experimental results	1.96	2.53	2.75	2.58	2.78	2.52	44.3481*
D. Methods, processes and procedures	1.72	2.40	2.28	2.18	2.96	2.21	51.1971*
E. Product, material and equipment informa-tion	1.87	2.27	2.41	2.09	2.35	2.19	27.2499*
F. Computer programs and model building information	0.80	1.38	1.50	1.91	2.43	1.52	83.2269*
G. Standard and patent specifications	2.08	1.91	2.58	1.05	1.17	1.49	73.4616*
H. Physical, technical and design data	2.29	2.62	2.46	2.34	2.43	2.42	13.2545
I. S and T news	2.66	1.13	2.73	2.70	2.91	2.74	22.7967
J. Statistical, economic business and general information	0.88	1.10	1.38	1.16	1.00	1.16	38.2992*

Key : *, The association is statistically significant at p = 0.05 as the tabulated value of x^2 at p = 0.05 for 16 degrees of freedom is 26,296; *Qualification* - U, Undergraduate; D, Diploma holder; B-Graduate; M, Post-graduate; P-Doctorate.

Note : The contigency tables on which chi-square test is made are not shown.

(r_s = 0.98), computer programs and model building information (F) (r_s = 0.72) and physical, technical and design data increases linearly. But the requirement of standard and patent specification (G) (r_s = –0.72) decreases linearly as status increases.

It may be seen in Table 5.4 that the requirement of all types of information except physical, technical and design data (H) and S&T news (I) has varied significantly with the level of qualifications of the space technologists. The relation is not linear in case of any type of information.

Table 5.5 shows that the need for state-of-the-art and review literature

TABLE 5.5 : Nature and Type of Information Required Vs Nature of Work (Mean Score)

Nature and Type of Information	Nature of work							Total	x^2
	A	B	C	P	D	E	F		
A. State-of- the- art	2.83	2.47	2.18	2.32	1.21	1.56	1.85	1.95	97.8563*
B. Tehoretical background	2.62	2.84	2.83	3.03	2.58	2.74	2.90	3.24	29.5012
C. Experimental results	2.43	2.44	2.71	2.89	2.40	2.53	2.79	2.52	27.2735
D. Methods, processes and procedures	2.20	2.20	2.29	2.50	2.04	2.31	2.05	2.20	45.5626*
E. Product, material and equipment information	2.45	1.79	2.31	2.41	2.15	2.19	1.89	2.19	22.2593
F. Computer programs and model building information	1.55	1.95	1.69	1.82	1.13	1.38	1.30	1.52	46.2171*
G. Standard and patent specifications	1.36	1.14	1.45	1.39	1.76	1.58	1.74	1.97	39.3603*
H. Physical, technical	2.24	2.16	2.57	2.68	2.43	2.10	2.16	2.41	35.7904
I. S and T news	2.83	2.72	2.74	2.61	2.76	2.62	2.50	2.73	21.9382
J. Statistical, economic, business and general information	1.43	1.43	1.16	0.08	0.98	1.26	1.25	1.16	41.4792*

Key : *, The association is statistically significant at $p = 0.05$ as the tabulated value of x^2 at $p = 0.05$ for 24 degrees of freedom is 36.415; *Nature of work* A, Management/Supervision; B, Planning/System Analysis; C, Design and Development; P, Both C & D; D, Fabrication and Testing; E, Operational Activity : F, Others.

Note :The contigency tables on which chi-square test is made are not shown

TABLE 5.6 : Nature and Type of Information Required Vs Specialisation of Users (Mean Score)

Nature and Type of Information	Specialisation						Total	x^2
	A	C	E	F	H	I		
A. State-of-the-art	2.60	1.74	1.17	2.89	2.06	2.21	1.94	83.9622*
B. Theoretical background	2.98	2.85	2.71	3.11	2.71	2.68	2.74	20.1062
C. Experimental results	2.67	1.74	2.34	2.42	2.71	2.70	2.54	39.5062*
D. Methods, processes and procedures	2.62	2.32	2.07	2.32	2.29	2.18	2.20	29.7875
E. Product, material and equipment information	2.45	1.11	2.23	1.84	2.24	2.26	2.20	43.0080*
F. Computer programs and model building information	2.00	3.10	0.85	2.05	1.53	1.51	1.46	104.3037*
G. Standard and patent specification	0.93	1.11	2.15	0.95	1.35	1.35	1.52	67.1592*
H. Physical, technical and design data	2.29	1.83	2,53	2.11	2.41	2.49	2.43	29.4793
I. S and T news	3.05	2.65	2.60	2.32	2.94	2.82	2.75	39.9819*
J. Statistical, economic, business and general information	1.21	1.22	0.92	0.63	1.24	1.19	1.09	32.1956*

Key : *, Association is statistically significant at $p = 0.05$ as the tabulated x^2 value at $df = 20$ and $p = 0.05$ is 31.410; *Specialisation*-A, Physics; C. Mathematics; E, Mechanical Engineering; F, Aeronautical and Structural Engineering; H, Electrical Engineering; I, Electronics.

Note : The contingency tables on which chi-square test is made are not shown.

(A), methods, processes and procedures (D), computer programs and model building information (F), standard and patent specifications (G) and statistical, economic, business and general information has varied significantly with the nature of work of the IST.

As could be seen from Table 5.6, physicists have sought more of state-of-the-art and review literature (A), experimental results (C), product, material and equipment information (E) and S & T news (I) than other types of information. Compared to others, mathematicians have naturally sought more of computer programs and model building information (F), and mechanical engineers have sought more of standard and patent specifications (G). The requirement of state-of-the-art and review literature (A) and theoretical background (B) is strongest among aeronautical and structural engineers. Electrical engineers require more of experimental results (C) and S & T news (I) than other types of information. Lastly, electronics engineers also have sought more experimental results (C) together with physical, technical and design data (H) than other specialists.[3]

It can also be inferred from the Pearson product moment correlation coefficients *(r)* in Table 5.2 that higher the experience of the space technologists higher the need for state-of-the-art literature (A) ($r = 0.16$), standard and patent specifications (G) ($r = 0.12$) and S & T news (I) ($r = 0.13$), but lower the need for theoretical background (B) ($r = -0.12$) and computer programs and model building information (F) ($r = -0.21$) and *vice versa.*

The professional activities and achievements of the IST are positively correlated with the need for state-of-the-art literature (A) ($r = 0.42$), methods, processes and procedures information (D) ($r = 0.13$), and computer programs and model building information (F) ($r = 0.23$) but negatively correlated with the need for standard and patent specifications (G) ($r = -0.21$).

The need for the state-of-the-art literature for the IST is, very highly, linearly and positively correlated/associated with status ($r_s = 0.95$), qualifications and nature of work, slightly and positively with experience ($r = 0.16$), moderately and positively with professional activities and achievements($r = 0.42$). The need for theoretical background is almost linearly and positively related to qualifications of users and slightly and negatively correlated ($r = -0.12$) with length of experience of users. The need for experimental results is very highly and positively correlated ($r_s = 0.92$) with status of users. The qualifications and specialisation of users are also related to the need for experimental results. The degree of requirement of information related tomethods, processes and procedures showed a significant relation with qualifications and nature of work and a slight positive correlation (r = 0.13) with professional activities and achievements of the users. The need for product, material and equipment information has almost perfectly and positively correlated (r_s=0.98) with status and has a significant relation with qualifications and specialisation of the users. Mathematicians and aeronautical and structural engineers have lesser need and others a higher need for product, material and equipment information which is on line with findings of Shuchman (1981, p. 68). The need for computer programs and model building information has a low negative correlation (r = –0.21) with experience, low positive correlation (r = 0.23) with professional activities and achievements of users, a high positive rank order correlation (r_s = 0.73) with status, a linear relationship with qualifications, curvilinear relationship with nature of work and a significant relationship with specialisation of the IST[4].

The need for standard and patent specifications is highly and negatively correlated with status (r_s = –0.72), slightly and positively with experience (r = 0.12), slightly and negatively with professional activities and achievements (r = –0.21), and has significant relation with qualifications, nature of work and specialisation of the users. The need for physical, technical and design data is highly and positively correlated (r_s=0.72) only with status of the users. The requirement of S & T news is significantly related to specialisation of the users and slightly and positively correlated (r = 0.13) with length of experience of the users. The 'need for statistical, economic, business and general information' is highly and positively correlated (r_s = 0.71) with status and significantly related to qualifications, nature of work and specialisation of the users.

From the above discussion and the data in Table 5.3, it is clear that the status of the user has a significant relation with all types of information listed, except S & T news and methods, processes and procedures. Hence, the observation of B.N.Singh (1981, p. 182) that status does not influence the information-needs is disproved in the present study. Further, the general observation of Garvey and others (1975, p. 501) that the information needs of least experienced are greater than those of the most experienced is true, as far as theoretical background/ basic S & T information (not S & T news) and computer programs and model building information are concerned. However, the reverse is true as far as state-of-the-art, S & T news and standard and patent specifications are concerned (See Table 5.2 under experience column).

In case of American industrial engineers, Shuchman (1981) has reported that out of the five variables used to evaluate the responses to nature and type of information sought, the job activity and type of industry appeared to be the two that make the most consistent difference. The qualification (degree) and the data of acquiring the qualification were not significant variables. As it could be seen from the Table 5.2, the present study does not support the observation of Shuchman. For the Indian space technologists, the status and qualifications made the most consistent difference in (eight of the ten) types of information sought followed by the subject of specialisation (in seven types of information), nature of work and experience (each in five types of information) and professional activities and achievements (in four types of information).

5.3 Effect of Change in Nature of Work on Information Requirements

A follow-up discussion held with selected respondents who had substantial change in the nature of work in recent years (Q. No.13.2) showed, broadly four types of changes in nature of work and their effect on information requirements and information-seeking activities; (i) Changing from operational activity, testing, design and development to supervision and planning has generally increased the need for information. In case of pure R&D works, the original activities continued and for additional responsibilities delegation of information-gathering work became inevitable; (ii) Changing from a specialised area to project work (*i.e.* from a more specific area to general area) has considerably decreased the need for information except routine information and vice versa. Delegation of information-gathering becomes a way of doing work in project, though the need for subject-information is reduced; (iii) Changing from a project work to a facility/service and production sector resulted in further reduction in need for information as the lack of any information did not drastically affect the work in facility/service and production sector. Lack of time and motivation are two major reasons for not seeking information in this area; (iv) Changing from engineering and technology-oriented work to slightly science-oriented work has resulted in increased need for information, particularly, theoretical background and basic S&T information.

5.4 Future Information Requirements

A free discussion with selected space technologists about their future information requirements revealed that a large majority (over 60%) do not foresee any significant change in their information requirements in the near future. Even if there is a little change they do not find it difficult to adapt to the circumstances over time. Out of the rest who thought of some changes in their information requirements in future, few expressed their fear at the problem of availability of increased quantity of information than what one could digest (*i.e.*, the problem of 'information overload'). This is not merely because of production of information, but because of enlargement ofboundary of interest and activities of the organisation. Others thought of the impact of electronics boom and some others explained how the organisation has to intensify its activities, increase its capabilities, reliability, quality and hence risks. The outcome of the discussion is that they are recognising the need for some condensed information services, new media of information and fast serving information agents. A sort of digesting agents as intermediary professionals are likely to come up in the core teams of project both to cut short delays and face the problem of information-overload.

NOTES

1. The percentages and other interpretations to the open ended category 'others' (K) have to be viewed with caution. Under responses to 'others', five were explicitly for 'information about educational apportunities in the country and abroad', three about 'management-techniques-related information', and two about 'internal/local procedural information like administrative and accounts procedures', three were closely related to category I, 3 to H, 2 to E and one each to A,C and D. The rest were vague and a conglomeration of more than one category.
2. The requirement of information on methods, processes and procedures not only interrelated moderately and positively ($r = 0.46$) with the requirement of product, material, equipment and apparatus information but also showed a low, but definite correlation ($r = 0.30$) with the purposes of designing and development of products and components as well as setting up and use of equipment. The requirement of product, material, equipment and apparatus information showed an almost moderate positive correlation ($r = 0.39$ and $r = 0.4$ respectively) with the above two purposes of seeking information.
3. However, the need for following types of information varied significantly with the subject of specialisation of the IST. State-of-the-art literature (A), experimental results (C), product, material and equipment information (E), computer programs and model building information (F), standard and patent specifications (G), S & T news (I) and statistical, economic, business and general information (J).
4. Shuchman (1981, p. 73-74) has reported that "The aerospace engineer is a large user of computer information".

6

MEANS OF ACQUIRING INFORMATION BY THE INDIAN SPACE TECHNOLOGISTS

6.1 Sources of Bibliographic (Reference) Information

Based on replies to the questionnaire (Q. No. 6), the sources of bibliographic information used by the space technologists is presented in Table 6.1. Some of the highly potential sources of bibliographic information for a majority (in the decreasing order of their dependence) are: library catalogue (69.3 per cent), colleagues and fellow professionals (67.5 per cent), and (browsing) library shelves (60.8 per cent). The other sources which are considered useful by nearly a half of the respondents are : (consulting) experts in the field (51.1 per cent), citations in the current reading materials (48.4 per cent), abstracting and indexing journals (46.6 per cent), bibliographies and review literature (45.5 per cent). To continue the above ranking of the sources, it is found that (consulting) library staff (40.2 per cent) and current awareness bulletins of the library (38.8 per cent) are the sources used marginally by less than half of the respondents. Browsing in bookshops (17.0 per cent) and announcements from publishers and booksellers (15.7 per cent) are regarded as sources by a very meagre number of the respondents[1].

From the data it can be concluded that consulting the library catalogue, browsing through library shelves, consulting colleagues, fellow professionals and experts, both within and outside the organisation and citations in current reading materials make the bulk of the sources of bibliographic information to the IST. Though there are quite a good number of bookshops in Bangalore and numerous catalogues and announcements of publishers and booksellers are displayed in the library

TABLE 6.1 : Sources of Bibliographic Information Used

Sources of Bibliographic Information	Not used		Used	
	No.	%	No.	%
A. Library catalogues	163	30.7	367	69.3
B. Abstracting and indexing periodicals/Journals	283	53.4	247	46.6
C. Bibliographies and review literature	288	54.5	240	45.5
D. Current awareness bulletins of the library	322	61.2	204	38.8
E. Citations in current reading materials	274	51.6	256	48.4
F. Announcements from publishers and booksellers	447	84.3	83	15.7
G. Browsing library shelves	208	39.2	322	60.8
H. Browsing in bookshops	440	83.0	90	17.0
I. Consulting experts in the field	259	48.9	271	51.1
J. Consulting colleagues and fellow professionals	172	32.5	358	67.5
K. Consulting library staff	317	59.8	213	40.2
L. Other means	—	—	13	2.5

Notes :
1. The number of respondents not answered to any category (except L) is to a maximum of 6 and invalid response a maximum of 4 and both are excluded.
2. The percentages of anonymous responses to the sources of bibliographic information used respectively are 72.4, 24.1, 37.9, 51.7, 42.9, 10.3, 65.5, 17.2, 41.4, 75.9, 13.8 and 3.6.

and distributed to the users, they are not found to be directly 'used' by/ 'useful' to a majority of them.

Except a higher dependence on libray catalogues, the dependence on rest of the sources of bibliographic information by the IST is in line with the findings of earlier studies. A high dependence on professional colleagues and experts found by Voigt (1959, p. 179-180), Hanson (1964, p. 71), a moderate dependence on citations in current reading materials found by Voigt (1959, p. 179-180), Raisig and others (1966) and B.N.Singh (1981, p. 172) are confirmed in the present study. Unusual higher dependence on library catalogues by the IST needs further study to explore the reasons. Yet as the data is based on opinion of users, it may be partly projective and partly due to inhibitions in the interpersonal communications which drive them away to impersonal sources like library catalogues.

As against 46.6 per cent of the IST depending on abstracting and indexing journals for bibliographic information, 38 per cent British physicists, 48 per cent American physicists (Gray, 1950, p. 417; Urquhart, 1965), 33.5 per cent American engineers (Davis, 1965, p. 31) and 14.9 per cent British mechanical engineers (Wood, 1967, p. 220) used indexing and abstracting journals in the respective studies. A slightly improved situation in case of the IST indeed.

The space technologists have felt that relevant references are rarely available in a ready to use way, but a lot of browsing, searching and digging are necessary. In the process they have often sought free reprints and reports and used indirect methods like cyclic way of generating relevant references. A respondent says 'searching in library shelves may not be scientific, but more practical'. The interdependence on different sources of bibliographic information is obvious from what an engineer has reported in the weekly summary sheet *i.e.*,

> "When I needed more information on 'Realising Switching Function', I went to the library and scanned through the latest books. I wondered that there are so many new techniques. I selected a latest book on 'Digital Design' which I have not yet read. A colleague suggested another book. When I searched in the library through author and subject indexes, I could not get more details of the book. I found it better to take some other book available on the shelf with the required information than searching through indexes".

A study (Sridhar, 1986) based on observations and discussions with the space technologists to know their habits and attitudes towards the card catalogues of ISAC library revealed that more than half of the use of card catalogues is to locate specific documents on shelf (*i.e.*, to get call number or report number), another one-fourth of the use is to interact with circulation system to find out the checked-out status of documents and possibly to reserve if checked-out, one-tenth of the use is for comprehensive subject search and the rest for miscellaneous purposes. The card catalogue consultations over a typical working day followed bimodal roughly symmetric distribution with maximum consultations around 10.30 hours and 15.30 hours like the pattern of user visit to library (Sridhar, 1982). The study has established the non-use of the classified catalogue for books and questioned the need for author and title catalogues for technical reports. The subject catalogue is extensively (54per cent) used by the space technologists.

6.2 Correlation of use of Sources of Bibliographic Information with User-characteristics

Results of chi-square statistic tests for association of sources of bibliographic information used with selected user-characteristics, namely, qualifications, nature of work, specialisation, experience and professional activities and achievements of the space technologists and rank order correlation with status are presented in Table 6.2. The Tables 6.4, 6.5, 6.6, 6.7, 6.8 and 6.3 present the data on sources of bibliographic information used against each one of the above user- characteristics in the same order.

It is clear from Tables 6.2 and 6.3 that higher the status, greater is the dependence on bibliographies and review literature ($r_s = 0.89$), current awareness bulletins ($r_s = 0.79$), citations in current reading materials ($r_s = 0.92$) and lesser is the dependence on library card catalogues ($r_s = -0.98$) and library staff ($r_s = -0.84$) for bibliographic references.

As could be seen from Table 6.4, lesser the qualifications of the space technologists, higher the consultation of library card catalogues, library staff and browsing in bookshops for bibliographic references and *vice versa.* Doctorates are twice likely to consult abstracting and indexing journals, bibliographies and review literature and citations in current reading materials for bibliographic information than others.

Consulting library catalogues, abstracting and indexing journals, library staff, browsing bookshops and library shelves and getting citations in current reading materials are significantly associated with the nature of work of the IST (Table 6.5). Increased with the management/ supervision content of the job, the IST have increasingly depended on bibliographies and reviews and decreasingly on library staff[2] for bibliographic references. Managers and supervisors have depended more on citations in current reading materials and planners and system analysts have depended more on browsing library shelves for bibliographic references than others.

The specialisation of the users is found to be significantly related to use of library catalogues, abstracting and indexing journals, bibliographies and review literature and announcements from publishers and booksellers (Table 6.6) for bibliographic information. Mechanical and electrical engineers depend significantly, more (mathematicians, the least) on library catalogues, electronics engineers (mechanical engineers, the least) on abstracting and indexing journals, physicists and aeronautical and

TABLE 6.2 : Results of Correlation/Association Tests of Sources of Bibliographic Information Used with (Selected) User-characteristics

	Sources of Bibliographic Information	Status	Qualifi-cations	Nature of work	Speciali-sation	Expe-rience	Professional Activities & Achieve-ments
		(r_s) $df = 8$	(x^2) $df = 4$	(x^2) $df = 6$	(x^2) $df = 5$	(x^2) $df = 5$	(x^2) Index $df = 5$
A	Library catalogues	-0.98*	24.44*	12.89*	14.59*	9.21	9.06
B	Abstracting and indexing journals	0.53	11.98*	8.91	16.24*	6.93	9.35
C	Bibliographies and review literature	0.89*	49.40*	14.76*	20.21*	6.15	17.53*
D	Current awareness bulletins of the library	0.79*	4.47	10.32	5.68	6.12	11.06
E	Citations in current reading materials	0.92*	34.33*	25.33*	8.30	17.01*	40.76*
F	Announcements from publishers and booksellers	-0.50	6.34	4.17	11.66*	3.50	6.96
G	Browsing library shelves	0.70*	2.13	12.91*	1.81	2.54	0.48
H	Browsing in bookshops	-0.52	10.44*	13.7*	2.64	3.76	10.05
I	Consulting experts	-0.39	12.78*	8.00	0.72	2.50	11.11*
J	Consulting colleagues	-0.28	5.47	3.24	3.97	10.45	6.05
K	Consulting library staff	-0.84*	45.48*	15.54*	8.19	2.08	22.38*

Key : *, The association/correlation is statistically significant at 0.05 significance level (p) for a two tailed test.

Note : The tabulated value of x^2 for a two tailed test at 0.05 significance level for 9,4,6 and 5 degrees of freedom are 16.919, 9.488, 12.592 and 11.070, respectively.

TABLE 6.3 : Percentages of Respondents Depending on Sources of Bibliographic Information Vs Status

Sources of Bibliographic Information	Status										Total	r_s
	A,B,C & D	E	F	G	H	I	J	K	L	M & N		
A Library catalogues	35.0	63.5	61.2	66.4	66.7	76.1	74.6	76.9	79.5	94.7	69.1	−0.9758*
B Abstracting and Indexing journals	55.0	63.5	53.7	52.3	44.4	37.3	32.8	26.9	61.5	42.1	47.5	0.5273
C Bibliographics and review literature	60.0	71.2	62.7	53.3	48.1	33.3	28.4	26.9	26.3	31.6	46.1	0.8909*
D Current awareness bulletins of the library	40.0	53.8	45.5	36.8	42.6	31.8	28.4	34.6	36.8	31.6	38.3	0.7909*
E Citations in current reading materials	90.0	69.2	62.7	56.1	46.3	38.8	31.3	26.9	30.8	31.6	48.8	0.9152*
F Announcements from publishers and booksellers	30.0	9.6	11.9	8.4	16.7	19.4	10.4	19.2	33.3	31.6	15.6	−0.5030
G Browsing library shelves	80.0	69.2	62.7	59.8	51.9	68.7	53.7	61.5	46.2	57.9	60.4	0.6970*
H Browsing in bookshops	30.0	5.8	13.4	9.3	22.2	17.9	16.4	26.9	28.2	31.6	16.8	−0.5212
I Consulting experts	35.0	42.3	49.3	43.9	51.9	65.7	56.7	65.4	51.3	42.1	51.0	−0.3939
J Consulting colleagues	55.0	57.7	61.2	74.8	75.9	71.6	65.7	61.5	69.2	57.9	67.4	−0.2848
K Consulting library staff	30.0	21.2	29.9	26.2	35.2	53.7	58.2	73.1	46.2	73.7	40.5	−0.8424*

Key : *, Correlations are statistically significant at $p = 0.05$ for 8 degrees of freedom *(df)* in a two tailed test.

Note : Status A to N are in the descending order of hierarchical rank.

TABLE 6.4 : Percentages of Respondents Depending on Sources of Bibliographic Information Vs Qualifications

	Sources of Bibliographic Information	Qualifications					Total	x^2
		U	D	B	M	P		
A	Library catalogues	83.9	79.2	66.0	58.4	56.5	69.1	24.4368*
B	Abstracting and indexing journals	41.9	40.6	47.7	49.7	78.3	47.3	11.9794*
C	Bibliographies and review literature	32.6	27.4	47.7	56.4	95.7	45.9	49.3977*
D	Current awareness bulletins of the library	32.6	40.0	44.4	35.1	40.9	38.6	4.4749
E	Citations in current reading materials	26.9	43.8	52.6	57.0	82.6	48.9	34.3250*
F	Announcements from publishers and booksellers	22.6	16.7	13.1	12.1	21.7	15.6	6.3391
G	Browsing library shelves	58.1	56.3	60.8	64.4	65.2	60.7	2.1337
H	Browsing in bookshops	26.9	15.6	16.3	12.1	8.7	16.5	10.4413*
I	Consulting experts	58.1	62.5	46.4	45.0	34.8	50.6	12.7824*
J	Consulting colleagues	60.2	72.9	70.6	67.1	56.5	67.5	5.4735
K	Consulting library staff	64.5	52.1	35.9	26.8	17.4	40.7	45.4826*

Key : * The association is statistically significant at $p = 0.05$ as the tabulated value of x^2 at $p = 0.05$ for 4 degrees of freedom is 9.488; *Qualifications*-U, Undergraduate; D, Diploma holder; B, Graduate; M, Postgraduate; P, Doctorate.

Note :The chi-square statistic test is done on actual number of respondents who checked as applicable or used (code 1) and *not* on percentage checked.

TABLE 6.5 : Percentages of Respondents Depending on Sources of Bibliographic Information Vs Nature of work

Sources of Bibliographic Information	Nature of work								
	A	B	C	P	I	E	F	Total	x^2
A Library catalogues	66.7	63.2	65.2	60.7	79.7	60.0	80.0	69.2	12.8905*
B Abstracting and indexing journals	50.0	49.1	49.1	35.7	44.2	34.3	70.0	46.8	8.9105
C Bibliographies and review literature	50.0	49.1	50.3	57.1	35.8	31.4	65.0	45.6	14.7640*
D Current awareness bulletins of the library	47.6	38.6	38.6	32.1	33.1	31.4	65.0	38.1	10.3175
E Citations in current reading materials	66.7	52.6	55.9	40.7	32.6	40.0	55.0	47.7	25.3333*
F Announcements from publishers and booksellers	16.7	12.3	12.4	14.3	20.3	14.3	15.0	15.4	4.1716
G Browsing library shelves	64.3	78.9	57.8	64.3	56.5	51.4	75.0	61.1	12.9145*
H Browsing in bookshops	21.4	12.3	11.2	17.9	19.6	17.1	40.0	16.6	13.6961*
I Consulting experts	50.0	38.6	49.1	46.4	58.7	48.6	60.0	50.9	8.0006
J Consulting colleagues	61.9	68.4	69.6	75.0	63.8	71.4	60.0	67.2	3.2431
K Consulting library staff	31.0	35.1	32.9	35.7	50.7	40.0	60.0	39.9	15.5396*

Key : *, The association is statistically significant at $p = 0.05$ as the tabulated chi-square value at $p = 0.05$ for 6 degrees of freedom is 12.592; *Nature of work* - A, Management/Supervision; B, Planning System analysis; C, Design and Development; P, Both C & D; D, Fabrication and Testing; E, Operational Activity; F, Others.

Note : The chi-square test is done on actual number of respondents who checked as applicable (code 1) and *not* on percentage checked.

structural engineers (mathematicians and electronics engineers the least) on bibliographies and reviews and physicists and electronics engineers on announcements from publishers and booksellers (aeronautical, structural and electrical engineers the least) for bibliographic references than others.

As per Table 6.7, as the experience of the space technologists increase, they increasingly depend on citations in current reading materials for bibliographic information.

Table 6.8 reveals that professionally more active space technologists are more likely to use bibliographies and review literature and citations in current reading materials and less likely to consult experts and library staff for bibliographic information. Those who use bibliographies/ reviews and citations in current reading materials have scored average professional activities and achievements index values of 12.8 and 13.0 respectively as against 7.1 and 5.4 of those not used. Surprisingly, those who do not consult library staff have higher (10.3) professional activities and achievements than those who consult them (8.9).

Consulting colleagues for bibliographic references is a common feature with all segments of users as it does not relate to any of the user-characteristics.[3] But consulting experts for references significantly decreases as the qualifications and professional activities and achievements increase.[4]

6.3 Delegation of Information-gathering Work

6.3.1 *The Degree of Delegation of Information-gathering Work*

The Indian Space Technologists were asked through the questionnaire (Q. No.7), whether they delegate the work of information-collection to others, the reasons for such delegation (as well as non delegation), and the nature of work delegated. Table 6.9 shows the number of respondents who delegate such work at three levels namely, occasionally, moderately and frequently and those who do not delegate. It is clear from the table that nearly 53 per cent of the respondents do not delegate and about 6 per cent delegate frequently (as against 72 per cent and 7 per cent respectively in INFROSS study vide Line, 1971, p. 425) the information-collection work. The remaining 41 per cent consists of those who delegate either moderately or occasionally. Compared to atomic scientists (66 per cent delegated sometimes, Hogg and Smith, 1959) and industrial technologists (64 per cent are willing to delegate, Scott, 1960, p. 33) of UK, the space technologists are willing to delegate to a lesser degree.

TABLE 6.5 : Percentages of Respondents Depending on Sources of Bibliographic Information Vs Specialisation

Sources of Bibliographic Information	Specialisation							
	A	C	E	F	H	I	Total	x^2
A Library catalogues	64.3	45.0	78.5	52.6	78.6	68.9	69.9	14.5868*
B Abstracting and indexing journals	45.2	35.0	33.3	47.4	42.9	54.5	45.8	16.2412*
C Bibliographies and review literature	69.1	45.0	37.3	73.7	57.1	43.4	45.8	20.2116*
D Current awareness bulletins of the library	31.7	30.0	33.6	47.4	50.0	42.5	38.8	5.6840
E Citations in current reading materials	54.8	50.0	39.3	57.9	64.3	51.8	48.9	8.2982
F Announcements from publishers and booksellers	19.1	5.0	11.9	5.3	7.1	22.1	16.8	11.6588*
G Browsing library shelves	61.9	75.0	61.5	57.9	64.3	60.4	61.5	1.8144
H Browsing in bookshops	14.3	20.0	19.3	15.8	7.1	14.4	15.9	2.6378
I Consulting experts	50.0	45.0	54.1	52.6	50.0	52.3	52.2	0.7158
J Consulting colleagues	64.3	65.0	63.0	73.7	71.4	72.1	68.4	3.9693
K Consulting library staff	31.0	35.0	44.4	15.8	42.9	43.2	40.9	8.1898

Key : *, Association is statistically significant as the tabulated value of chi-square at $p = 0.05$ for $df = 5$ is 11.070; *Specialisation* - A, Physics; C, Mathematics; E, Mechanical Engineering; F, Aeronautical and Structural Engineerings; H, Electrical Engineering; I, Electronics.

Note : The chi-square test is done on actual number of respondents who checked as applicable (code 1) and *not* on percentage checked.

TABLE 6.7 : Sources of Bibliographic Information Used Vs Experience

	Sources of Bibliographic Information	Average experience (in years) of those Used	Not used	x^2
A	Library catalogues	8.56	9.09	9.2094
B	Abstracting and indexing journals	9.12	8.38	6.9335
C	Bibliographies and review literature	8.97	8.52	6.1495
D	Current awareness bulletins of the library	9.23	8.40	6.1222
E	Citations in current reading materials	9.44	7.98	17.0087*
F	Announcements from publishers and booksellers	8.80	8.72	3.5049
G	Browsing library shelves	8.98	8.09	2.5448
H	Browsing in bookshops	8.80	8.71	2.7618
I	Consulting experts	8.68	8.78	2.5027
J	Consulting colleagues	8.18	9.87	10.4522
K	Consulting library staff	9.21	8.45	2.0761

Key : *, The association is statistically significant at 0.05 significance level (p) as the tabulated value of chi-square at 0.05 significance level for 5 degrees of freedom *(df)* for a two tailed test is 11.070.

Note : Chi-square test is made on six ranges of experience, namely, 00, 01–03, 04–06, 07–11, 12–16,17 and above.

6.3.2 *Correlation of Degree of Delegation of Information-gathering work with User-characteristics*

The non-parametric chi-square statistic test for association of degree of delegation of information-gathering work with the six user-characteristics shows significant association of degree of delegation with status, qualifications, nature of work and professional activities and achievements as shown in Table 6.10. But the experience is not related to the degree of delegation and hence is contrary to the finding of INFROSS study (Line 1971, p. 425). The Tables 6.11, 6.12, 6.13, 6.14, 6.15 and 6.16 present the contingency tables on which chi-square tests have been made. Respondents with higher status and higher qualifications tend to delegate more of information-gathering work as found in INFROSS study (Line, 1971, p. 425). Even in case of nature of work (Table 6.13), those with management, planning, design and development jobs tend to delegate more than others. However, the respondents with higher professional activities and

TABLE 6.8 : Sources of Bibliographic Information Used Vs Professional Activities and Achievements Index

	Sources of Bibliographic Information	Average index values of those Used	Not Used	x^2
A	Library catalogues	8.66	11.98	9.0593
B	Abstracting and indexing journals	10.65	8.99	9.3518
C	Bibliographies and review literature	12.78	7.12	17.5282*
D	Current awareness bulletins of the library	12.26	7.83	11.0581
E	Citations in current reading materials	13.03	5.42	40.7589*
F	Announcements from publishers and booksellers	12.46	9.26	6.9624
G	Browsing library shelves	9.84	9.61	0.4754
H	Browsing in bookshops	7.32	10.28	10.0473
I	Consulting experts	7.74	11.88	11.1148*
J	Counsulting colleagues	9.49	10.23	6.0522
K	Consulting library staff	8.90	10.25	22.3806*

Key : *, The association is statistically significant at 0.05 significance level (p) as the tabulated value of chi-square at 0.05 significance level for 5 degrees of freedom *(df)* for a two tailed test is 11.070.

Note : Chi-square test is made on six ranges of professional activities and achievements index, namely, 00, 01–05 06–10, 11–20, 21–30, 31 or more.

achievements (19.7) tend to delegate more of information-gathering work than lower professional activities and achievements (Table 6.16). Though not statistically significant (Table 6.14), pure science personnel (i.e., Physicists and Mathematicians) tend to delegate more than engineering personnel which is in conformity with earlier findings (Vagianos, 1971, p. 86).[5]

6.3.3 *Reasons for Delegating Information-gathering Work*

Nearly 74 per cent of respondents have opined that 'making team or group members involved' is the reason for delegating information-gathering work (see Table 6.17), about 38 per cent due to lack of time and another 20 per cent for unfamiliarity with the sources. A very small segment consisting of 9 per cent of the respondents felt that the reason could also

TABLE 6.9 : Degree of Delegation of Information-gathering Work

		No.	%
A	No Delegation	262	52.9
B	Delegate occasionally	103	20.8
C	Delegate moderately	101	20.4
D	Delegate frequently	29	5.9
	Total	495	100.0

Note : 1. Eight not answered and 32 invalid responses are excluded.

2. The anonymous responses to A,B,C and D respectively are 15,6,6 and 2.

TABLE 6.10: Results of Chi-square Test for Association of Degree of Delegation of Information-Gathering Work with (Selected) User-characteristics

User-characteristics	X^2	*df*	Tabulated value of x^2 at p = 0.05 for a two-talied test
Status	75.73*	21	32.671
Qualifications	53.36*	12	21.026
Nature of Work	39.83*	18	28.869
Specialisation	16.48	15	24.996
Experience	13.65	12	21.026
Professional Activities and Achievements Index	25.75*	6	12.592

Key : *, The association is statistically significant at 0.05 significance level.

be that the task is tedious and monotonous.[6] These findings appear to be unique to the IST.

6.3.4 *Correlation of Reasons for Delegating Information-gathering Work with User-characteristics*

Table 6.18 presents a summary of results of association tests between reasons for delegating information-gathering work and the user-characteristics. The contingency tables on which chi-square tests are

TABLE 6.11 : Degree of Delegation of Information-Gathering Work Vs Status

Status	No delegation	Degree of Delegation			
		Occasional	Moderate	Frequent	Total
A,B,C & D	1	9	7	3	20
E	17	17	12	6	52
F	29	19	15	4	67
G	50	23	26	4	103
H	25	12	11	4	52
I	35	10	11	6	62
J	36	9	13	1	59
K,L,M, & N	59	3	5	0	67
Total	252	102	100	28	482

($x^2 = 75.7296$, $df = 21$, $p < 0.05$)

Note : The status A to N are in the descending order of the hierarchical rank.

TABLE 6.12 : Degree of Delegation of Information-Gathering Work Vs Qualifications

Qualifications		No Delegation	Degree of Delegation			
			Occasional	Moderate	Frequent	Total
U	Undergraduate	63	3	6	1	73
D	Diploma holder	47	17	19	5	88
B	Graduate	68	37	31	11	147
M	Post graduate	62	42	36	8	148
P	Doctorate	12	1	6	4	23
	Total	252	100	98	29	479

(x^2 = 53.363, df = 12, $p < 0.05$)

made for finding association/correlation of reasons for delegating information-gathering work with user-characteristics are shown in Tables 6.19, 6.20, 6.21, 6.22, 6.23 and 6.24.

The reason of unfamiliarity with the sources of information for delegating information-gathering work has significantly varied negatively

TABLE 6.13 : Degree of Delegation of Information-Gathering Work Vs Nature of Work

	Nature of Work	No Delegation	Degrees of Delegation Occasional	Moderate	Frequent	Total
A	Management/Supervision	12	8	15	7	42
B	Planning/System Analysis	28	13	14	2	57
C	Design and Development	82	32	32	8	154
P	Both C and D	8	9	5	1	23
D	Fabrication and Testing	78	23	16	7	124
E	Operational activity	23	4	5	0	32
F	Others	13	1	3	2	19
	Total	244	90	90	27	451

($x^2 = 39.8251$, $df = 18$, $p < 0.05$)

TABLE 6.14: Degree of Delegation of Information-Gathering Work Vs Specialisation

	Specialisation	No Delegation	Degrees of Delegation Occasional	Moderate	Frequent	Total
A	Physics	19	9	10	4	42
C	Mathametics	7	6	4	2	19
E	Mechanical Engg.	72	24	19	3	118
F	Aeronautical and Structural Engg.	12	3	3	1	19
H	Electrical Engg.	6	2	7	1	16
I	Electronics	108	48	38	14	208
	Total	224	92	81	25	422

($x^2 = 16.4785$, $df = 15$, $p = 0.05$)

with status ($r_s = -0.93$) (Table 6.15), level of qualifications (Table 6.20) and professional activities and achievements (Table 6.24) of the space technologists. The only other reason which is significantly related to user-characteristics is involving team members which varied positively with status ($r_s = 0.9$), qualification level, management and supervisory nature of the job (Table 6.21) and length of experience (Table 6.23). The familiarity of source of information has increased with qualifications, status and professional activities. Interestingly, no doctorate has attributed unfamiliarity with source as a reason for delegating information-gathering work.

TABLE 6.15 : Degree of Information-gathering Work Vs Experience

		Experience in years					Total	Mean
	Degree of Delegation	**< 3**	**3–5**	**6–10**	**11–15**	**>15**		
A	No Delegation	40	31	61	43	14	189	8.11
B	Occasional Delegation	5	13	27	25	8	78	10.03
C	Moderate Delegation	15	15	26	18	9	83	8.72
D	Frequent Delegation	2	5	7	7	4	25	10.44
	Total	62	64	121	93	35	375	8.8

($x^2 = 13.6513$, $df = 12$, $p > 0.05$)

TABLE 6.16: Degree of Delegation of Information-gathering Work Vs Professional Activities and Achievements Index

	Degree of Delegation	Professional Activities and Achievements Index			Total	Mean
		0-5	**6-15**	**>15**		
A	No Delegation	59	27	9	95	8.09
B	Occasional Delegation	15	16	10	41	14.15
C	Moderate Delegation	13	11	15	39	18.22
D	Frequent Delegation	2	6	5	13	19.65
	Total	89	60	39	188	12.31

($x^2 = 25.7521$, $df = 6$, $p < 0.05$)

6.3.5 *Nature of Information-gathering work Delegated*

A maximum of 55.9 per cent of the space technologists delegate collection of factual, design and physical data[7] followed by current search for latest references (48.6 per cent), documents/literature collection (45 per cent), retrospective and exhaustive search[8] for references (26.6 per cent) and collection of analysed and digested literature/information (34.2 per cent) (See Table 6.25).

As found in earlier studies (Line, 1971, p. 425 and Raitt, 1984, p. 301), the space technologists delegate more of factual data collection (60 per cent) than literature search (45 per cent) but peculiarly delegate more of current search for latest references (48.6 per cent) than retrospective and exhaustive search (26.6 per cent) or analysed and digested literature (34.2 per cent). The latter difference is attributable to the fact that keeping

TABLE 6.17 : Reasons for Delegating Information-gathering Work

	Reasons	Not Applicable		Applicable	
		No.	%	No.	%
A	Lack of time	139	62.1	85	37.9
B	The task in tedious and monotonous	203	90.6	21	9.4
C	Unfamiliarity with the sources	179	79.9	45	20.1
D	Making team or group members involved	59	26.3	165	73.7
E	Other reasons	—	—	12	5.4

Note: 1. Those not answered and invalid replies are excluded from the table.
2. The reasons A to D have been checked respectively by 4,2,1 and 9 (*i.e.* 36.4%, 18.2%, 9.1% and 81.8%) anonymous respondents out of 11 anonymous respondents.

TABLE 6.18 : Results of Association Tests of Reasons for Delegating Information-Gathering work with (Selected) User-characteristics

User-characteristics	Degrees of Freedom	Reasons for Delegation			
		A	B	C	D
Status (r_S)	6	0.21	–0.23	–0.93*	0.90*
Qualifications (x^2)	4	1.47	3.02	11.94*	21.42*
Nature of work (x^2)	6	7.17	6.53	10.39	13.50*
Specialisation (x^2)	5	6.34	5.58	11.02	5.25
Experience (x^2)	5	4.63	6.04	10.77	12.39*
Professional Activities and Achievements Index (X^2)	5	8.03	4.60	15.06*	9.41

Key : *, The association/correlation is statistically significant at 0.05 significance level for a two tailed test.

Reasons for Delegation : A, Lack of time; B, Task is tedious and monotonous; C, Unfamilarity with the sources; D, Involving team members.

Note : The tabulated values of X^2 at 0.05 significance level for a two tailed test for 4, 5 and 6 degrees of freedom are respectively 9.488, 11.670 and 12.592.

TABLE 6.19 : Percentages of Respondents checking the Reasons for Delegating Information-Gathering Work Vs Status

	Reasons for Delegation	Status								Total	r_s
		A B C & D	E	F	G	H	I	J	K L M & N		
A	Lack of time	52.6	54.3	28.9	30.8	25.0	38.5	40.9	50.0	37.9	0.2143
B	Task is tedious and monotonous	10.5	8.6	7.9	7.7	8.3	7.7	18.2	12.5	9.4	–0.2321
C	Unfamiliarity with the sources	0.0	11.4	2.6	19.2	33.3	26.9	50.0	50.0	20.1	–0.9345*
D	Involving team members	89.5	91.4	97.4	71.2	62.5	57.7	40.9	37.5	73.7	0.9048*

Key : *, The correlation is statistically significant at $p = 0.05$ as per a two tailed t = test $df = 6$

Note : The status from A to N are in the descending order of hierarchical rank. The contingency tables on which chi-square test is carried out are not shown.

TABLE 6.20 : Percentages of Respondents checking the Reasons for Delegating Information-Gathering Work Vs Qualifications

Reasons for Delegation		Qualifications						
		U	D	B	M	P	Total	x^2
A	Lack of time	30.0	35.9	36.0	43.4	36.4	38.5	1.4679
B	Task is tedious and monotonous	40.0	12.8	9.3	6.0	9.1	9.2	3.0196
C	Unfamiliarity with the sources	40.0	33.3	22.7	13.3	0.0	20.6	11.9375*
D	Involving team members	40.0	53.8	72.0	85.5	90.9	73.4	21.4183*

Key : *, The association is statistically significant at p=0.05 as the tabulated value of x^2 at p = 0.05 at 4 degrees of freedcm is 9.488; *Qualifications*; U, Undergraduate; D, Diploma holder; B, Graduate, M, Postgraduate; P, Doctorate.

Note : The chi-square test is carried out on contingency tables of number of respondents checked and *not* on percentage of respondents checked.

TABLE 6.21 : Percentages of Respondents checking the Reasons for Delegating Information-Gathering Work Vs Nature of Work

Reasons for Delegation	Nature of work							Total	x^2
	A	B	C	P	D	E	F		
A Lack of time	50.0	41.4	30.0	28.6	34.9	50.0	66.7	36.5	7.1733
B Task is tedious and monotonous	10.0	3.5	5.7	14.3	16.3	0.0	16.7	9.0	6.5320
C Unfamiliarity with the sources	6.7	17.2	22.9	14.3	34.9	37.5	16.7	22.0	10.3874
D Involving team members	90.0	82.8	75.7	71.4	69.8	50.0	33.3	75.0	13.4954*

Key : *, Statistically significant association at p=0.05 as the tabulated value of x^2 at $p = 0.05$ for $df = 6$ is 12.592; *Nature of work* - A, Management/Supervision; B, Planning/System Analysis; C, Design and Development; P, Both C & D; D, Fabrication and Testing; E, Operational Activity; F, Others.

Note : The chi-square statistic test is done on contingency tables of number of respondents checked and *not* on percentage of respondents checked.

TABLE 6.22: Percentages of Respondents checking the Reasons for Delegating Information Gathering Work Vs Specialisation

	Reason for Delegation	Specialisation						Total	x^2
		A	C	E	F	H	I		
A	Lack of time	54.6	45.5	34.9	42.9	10.0	38.4	38.5	6.3388
B	Task is tedious and monotonous	18.2	0.0	11.6	0.0	0.0	8.1	8.9	5.5751
C	Unfamiliarity with the sources	18.2	9.1	14.0	0.0	0.0	29.3	20.8	11.0168
D	Involving team members	77.3	72.7	79.1	85.7	100.0	70.7	75.5	5.2516

Key : Specialisation-A, Physics; C, Mathematics; E, Mechanical Engineering, F, Aeronautical and Structural Engineering; H, Electrical Engineering; I, Electronics.

Notes : 1. There is no statistically significant association as the tabulated value X^2 at p=0.05 for df=5 is 11.670.

2. The non-parametric chi-square statistic test is done on contingency tables of number of respondents checked and not percentage of respondents checked.

TABLE 6.23 : Reasons for Delegating Information-Gathering Work Vs Experience

	Reasons for delegation	Average experience (in years) of those		
		Checked	Not checked	x^2
A	Lack of time	9.72	9.07	4.6256
B	Task is tedious and monotonous	10.93	9.19	6.0425
C	Unfamiliarity with the sources	6.97	9.86	10.7693
D	Involving Team Members	10.08	6.94	12.3884*

Key : *, The association is statistically significant at 0.05 significance level for 5 degrees of freedom *(df)* in a two tailed test as the tabulated critical value of x^2 is 11.670.

Note : The chi-square statistic test is done on contingency tables with the ranges of 00. 01–03, 04–06, 07–11, 12–16, 17 or more years of experience.

TABLE 6.24 : Reasons for Delegating Information-Gathering Work Vs Professional Activities and Achievements Index

	Reasons for Delegation	Average Professional Activities and Achievements Index value of those		
		Checked	Not checked	x^2
A	Lack of time	15.28	12.45	8.0273
B	Task is tedious and monotonous	16.14	13.55	4.5974
C	Unfamiliarity with the sources	10.53	14.39	15.0628*
D	Involving team members	15.32	8.23	9.4052

Key : *, The association is statistically significant at 0.05 significance level for 5 degrees of freedom *(df)* in a two tailed test as the tabulated critical value of X^2 is 11.070.

Notes : 1. The chi-square statistic test is made on six ranges of professional activities and achievements index values, namely, 00, 01–05, 06–10, 11–20, 21–30, 31 or more.

2. Though the average professional activities and achievements index values of those checked and not checked the reason D are respectively 15.32 and 8.23, the chi-square statistic is less than critical due to the fact that only 20 respondents did not check the reason as against 71 who checked it.

TABLE 6.25 : Nature of Information-Gathering Work Delegated

Nature of work		Not applicable		Applicable	
		No.	%	No.	%
A.	Retrospective and exhaustive search for references	163	73.4	59	26.6
B.	Current search for latest references	114	51.4	108	48.6
C.	Factual data such as design data and physical data	98	44.1	124	55.9
D.	Documents/Literature	122	55.0	100	45.0
E.	Analysed and Digested literature/information	146	65.8	76	34.2
F.	Others	—	—	3	1.4

Notes : 1. Not answered and invalid responses are excluded.

2. The percentages of anonymous responses checking A to F are respectively 8.3, 41.7, 58.3, 58.3, 16.7 and 0.

TABLE 6.26 : Results of Association/Correlation Tests of Nature of Information-Gathering Work Delegated with User-Characteristics

User-characteristics	Degrees of Freedom	Nature of Information-Gathering Work Delegated				
		A	B	C	D	E
Status (r_s)	6	0.38	0.38	0.55	0.26	–0.48
Qualifications (x^2)	4	3.69	1.58	3.37	10.06*	8.18
Nature of Work (x^2)	6	6.86	10.64	16.19*	15.14*	3.68
Specialisation (x^2)	5	3.13	6.55	8.91	4.67	8.78
Experience (x^2)	5	17.07*	10.40	10.42	7.28	2.62
Professional Activities and Achievements Index (x^2)	5	13.60*	11.43*	8.97	3.12	6.03

Key : *,The association/correlation is statistically significant at 0.05 significance level; *Nature of Information-Gathering Work Delegated*-A, Retrospective and exhaustive search for references; B, Current search for latest references; C, Factual data; D, Documents/Literature ; E, Analysed and digested information.

Note : The tabulated value of x^2 at 0.05 significance level for a two tailed test for 4,5 and 6 degrees of freedom are respectively 9.488, 11.670 and 12592.

up-to-date is their major purpose of seeking information.

6.3.6 *Correlation of Nature of Information-gathering Work Delegated with User-characteristics*

The results of association test of nature of information-gathering work delegated with user-characteristics are depicted in Table 6.26. The individual contingency tables are provided in Tables 6.27, 6.28, 6.29, 6.30, 6.31 and 6.32 respectively for status, qualifications, nature of work, specialisation, experience and professional activities and achievements. Delegation of search for document/literature has varied significantly with level of qualifications (Table 6.28) and nature of work (Table 6.29) of the space technologists. Postgraduates with management/supervisory, planning/system analysis work and undergraduates with operational activities have delegated collection of documents/ literature than others. Doctorates are 50 per cent less likely to delegate selection and collection of documents/literature than others. Increasingly, more of retrospective and exhaustive search for references is delegated by the space technologists with longer experience (Table 6.31) and higher professional activities and achievements (Table 6.32). Increased professional activities and achievements is likely to cause increased delegation of even current search for latest references (Table 6.32). The nature of work exhibited curvilinear relation with delegation of documents/literature-collection (Table 7.29). Obviously managers/supervisors tend to delegate literature- collection to involve group members and operational activity personnel due to unfamiliarity with the sources of information.

6.3.7 *The Reasons for Non-delegation of Information-gathering Work*

Two reasons for non-delegation of information-gathering work checked by largest number of respondents are, not having assistants (42.2 per cent) and the importance of information-gathering/searching process itself *i.e.*, the serendipity value of information-searching (40.3 per cent) (See Table 6.33). The remaining reasons in the table are checked by a few respondents.[9] A respondent says "while searching there are ample chances for me to sight other titles which will be of immense help in my field". By and large, many of the space technologists felt that communicating their information-needs to others for collecting information on their behalf and training others in their ways of collecting information are not difficult tasks and also believed that others can collect information for them, though a

TABLE 6.27 : Percentages of Respondents checking Nature of Information-Gathering work Delgated Vs Status

	Nature of Information-Gathering Work Delegated	Status								Total	r_s
		A B C & D	E	F	G	H	I	J	K L M & N		
A	Retrospective and exhaustive Search for references	57.9	37.1	21.6	22.0	20.8	14.8	22.7	28.6	26.7	0.3810
B	Current search for latest references	47.4	68.6	45.9	40.0	50.0	48.2	40.9	42.9	48.4	0.3810
C	Factual data	63.2	77.1	59.5	46.0	41.7	48.1	59.1	57.1	56.1	0.5476
D	Documents/Literature	47.4	60.0	54.1	50.0	20.8	33.3	31.8	57.1	45.2	0.2619
E	Analysed and digested information	26.3	40.0	35.1	22.0	25.0	51.9	36.4	71.4	34.4	–0.4762

Key : *, The correlation is statistically significant at $p = 0.05$ as per a two tailed t-test with $df = 6$.

Note : The status from A to N are in the descending order of hierachical rank. The contingency tables on which chi-square test is carried out are not shown.

TABLE 6.28 : Percentages of Respondents checking Nature of Information-Gathering work Delegated Vs Qualifications

Nature of Information-Gathering Work Delegated		Qualifications U	D	B	M	P	Total	x^2
A	Retrospective and exhaustive search for references	88.9	18.0	29.3	29.3	36.4	26.9	3.6948
B	Current search for latest references	44.4	51.3	49.3	45.1	63.6	48.6	1.5833
C	Factual data	44.4	66.7	52.0	52.4	63.6	55.1	3.3720
D	Documents/Literature	66.7	33.3	40.0	56.1	27.3	45.4	10.0600*
E	Analysed and digested information	55.6	43.6	24.0	40.2	27.3	35.2	8.1822

Key : *, The association is statistically significant at $p = 0.05$ as the tabulated value of x^2 at $p = 0.05$ for 4 degrees of freedom is 9.488;
Qualifications-U, Undergraduate; D, Diploma holder; B, Graduate; M, Postgraduate; P, Doctorate.

Note : The chi-square test is carried out on contengency tables of number of respondents checked and *not* on percentages of respondents checked.

TABLE 6.29 : Percentages of Respondents Checking Nature of Information-gathering Work Delegated Vs Nature of Work (of User who Delegated)

Nature of Information Gathering Work Delegated	Nature of Work (of Users)								
	A	B	C	P	D	E	F	Total	x^2
A Retrospective and exhaustive search for references	40.0	35.7	24.6	21.4	20.9	22.2	0.0	26.8	6.8562
B Current search for latest references	56.7	60.7	42.0	64.3	44.2	11.1	60.0	48.0	10.6355
C Factual data	80.0	42.9	55.1	7.1	44.2	33.3	60.0	55.6	16.1898*
D Documents / Literature	66.7	60.7	34.8	28.6	34.9	55.6	40.0	43.9	15.1364*
E Analysed and digested information	30.0	28.6	39.1	21.4	41.9	44.4	40.0	35.9	3.6819

Key : *,The association is statistically significant at $p = 0.05$ as the tabulated value of x^2 for $df = 6$ is 12.592. *Nature of work* - A, Management/Supervision; B, Planning System Analysis; C, Design and Development; P, Those checked both C & D; D, Fabrication and Testing E, Operational Activity; F, Others.

Note : The chi-square statistic test is done on contingency tables of number of respondents checked and *not* on percentage of respondents checked.

TABLE 6.30 : Percentages of Respondents checking Nature of Information-Gathering Work Delegated Vs Specialisation

Nature of Information Gathering Work Delegated	Specialisation							
	A	C	E	F	H	I	Total	x^2
A Retrospective and Exhaustive Search for references	30.4	27.3	16.3	28.6	30.0	29.9	26.7	3.1263
B Current search for latest references	47.8	36.4	34.9	28.6	70.0	50.5	46.1	6.5503
C Factual data	60.9	36.4	72.1	57.1	30.0	58.8	59.2	8.9082
D Documents/Literature	39.1	63.6	48.8	42.9	20.0	44.3	44.5	4.6662
D Analysed and digested Information	26.1	45.5	25.6	42.9	0.0	38.1	32.5	8.7821

Key : *Specialisation*-A, Physics; C-Mathematics; E-Mechanical Engineering; F-Aeronautical and Structural Engineering; H-Electrical Engineering; I-Electronics.

Notes : 1. There is no statistically significant association as the tabulated value of x^2 at $p = 0.05$ for $df = 5$ is 11.670.

2. Chi-square test is done on contingency tables of number of respondents checked and *not* on percentage of respondents checked.

TABLE 6.31 : Nature of Information-Gathering Work Delegated Vs Experience

Nature of Information-Gathering Work Delegated	Average Experience of those Checked	Not checked	x^2
A Retrospective and exhaustive search for references	10.17	8.90	17.0733*
B Current search for latest references	10.31	8.32	10.3965
C Factual data	10.11	8.14	10.4205
D Documents / Literature	10.02	8.64	7.2803
E Analysed and digested information	8.86	9.49	2.6192

Key : *, The association is statistically significant at 0.05 significance level for 5 degrees of freedom *(df)* in a two tailed test as the tabulated critical value of x^2 is 11.670.

Note : The chi-square test is done on contingency tables with the ranges of 00, 01–03, 04, 06, 07–11, 12–16, 17 or more years of experience.

TABLE 6.32 : Nature of Information-Gathering Work Delegated Vs Professional Activities and Achievements Index

Nature of Information-Gathering Work Delegated	Average Professional Activities and Achievements Index value of those Checked	Not checked	x^2
A Retrospective and exhaustive search for references	18.59	11.77	13.6005*
B Current search for latest references	16.47	11.51	11.4297*
C Factual data	15.41	11.74	8.9736
D Documents / Literature	13.67	13.86	3.1241
E Analysed and digested information	14.57	13.33	6.0265

Key : *, The association is statistically significant at 0.05 significance level for 5 degrees of freedom *(df)* in a two tailed test as the tabulated critical value of x^2 is 11.670.

Note : The chi-square statistic test is made on six ranges of professional activities and achievements index values, namely, 00. 01–05, 6–10, 11–20, 21–30, 31 or more.

TABLE 6.33 : Reasons for Nondelegation of Information-Gathering Work

Reasons	Not applicable		Applicable	
	No.	%	No.	%
A. Do not have assistants to delegate	170	57.8	124	42.2
B. Find it difficult to communicate my information needs to others	266	90.8	27	9.2
C. Do not believe in others collecting information for me	264	90.1	29	9.9
D. Find it difficult to train others in my ways of collecting information	261	89.1	32	10.9
E. Do not want to disturb others	235	80.2	58	19.8
F. Believe that the searching process is as important as information itself	175	59.7	418	40.3
G. Others reasons	-	-	18	6.1

Note : 1. Two hundred and four not answered and 2 invalid responses are excluded.

2. The percentages of anonymous respondents who checked the reasons A to G are 42.9, 7.1, 7.1, 0, 42.9, 57.1 and 7.1 respectively.

minority said that there are not many people in the organisation with similar background to delegate. Both serendipity value of information-searching process and not having assistants to delegate are unique reasons for non-delegation found with the space technologists. The problems of need for empathy between delegator and delegatee (Herner and Herner, 1967, p. 28) and difficulty of communicating requirements to others (Myers, 1970, p. 27) found earlier are less serious with the space technologists

6.3.8 *Correlation of Reasons for Nondelegation of Information-gathering Work with User-characteristics*

Table 6.34 presents the summary results of association/correlation tests of reasons for non-delegation of information-gathering work with user-characteristics. The individual contingency Tables of association of reasons for non-delegation with status, qualifications, nature of work, specialisation, experience and professional activities and achievements are shown in Tables 6.35, 6.36, 6.37, 6.38, 6.39 and 6.40 respectively.

The difficulty of training others in one's ways of gathering information (r_s = 0.95) and the importance of searching process due to serendipity values (due to accidental acquisition of information and other incidental benefits of browsing and searching) (r_s = 0.92) have increased with increase in status of users (Table 6.35). The belief in serendipity value of information searching process has also increased as the level of education increases (Table 6.36). The reasons of not having assistants, not willing to disturb others and serendipity value of information-gathering work are found to be significantly varying with nature of work of the space technologists (Table 6.37). Professionally active users (index 9.8) are twice likely to believe in the serendipity value of information-searching process than others (index 4.9) (Table 6.40).

By and large, delegation of information-gathering work is done by a minority of the space technologists, mainly to make the team members involved or as part of a team work. There exists a sort of division of labour among members of the team so that one or two of them will do the information-gathering work. Obviously, the space technologists with higher status, higher qualifications, higher managerial content of the job and higher professional activities tended to delegate more. Apart from making team members involved, the second major reason for delegating information-gathering work is lack of time. In addition, persons with lower status and qualifications depend on others due to unfamiliarity with the sources of information. Expectantly, a majority (55.9per cent) of the space technologists, especially managers and supervisors, have delegated collection of factual, physical and design data. Though the task of retrospective and exhaustive search for references is tedious, it is the least delegated information-gathering activity. One of the reasons attributable for such a finding is that the need for retrospective and exhaustive search among the space technologists occurs rarely. But such retrospective searches, whenever needed, are delegated more by those with longer experience and higher professional activities and achievements than others. However, certain works such as search for latest references, collection of relevant literature and preparation of digest on a topic have been moderately delegated.

6.4 Accidental Acquisition of Information

When accidental acquisition of information was discussed during interviews, many interviewees felt that it occurred rarely to them. Yet very

TABLE 6.34 : Results of Association/Correlation Tests of Reasons for Nondelegation of Information-Gathering Work With (Selected) User-Characteristics

	Reasons for Non-delegation	Status (r_s) $df = 6$	Qualifications (x^2) $df = 4$	Nature of work (x^2) $df = 6$	Specialisation (x^2) $df = 5$	Experience (x^2) $df = 5$	Professional Activities and Achievements Index (x^2) $df = 5$
A	No assistants	–0.48	9.28	12.87*	2.85	7.64	5.53
B	Difficulty in communicating needs	–0.07	3.35	5.93	3.64	8.06	10.73
C	Do not believe in others collecting information	–0.05	1.36	2.21	2.22	5.17	2.21
D	Difficult to train others	0.95*	8.67	8.19	5.94	2.67	5.59
E	Do not want to disturb others	0.21	4.13	13.03*	4.06	2.01	5.09
F	Searching process is as important as information itself	0.92*	34.13*	21.38*	8.54	7.47	11.89*

Key : *, The association/correlation is statistically significant at 0.05 significance level (p).

Note : The tabulated values of x^2 at 0.05 significance level for a two tailed test for 4,5 and 6 degrees of freedom (*df*) are respectively 9.488, 11.670 and 12.592.

TABLE 6.35 : Reasons for Nondelegation of Information-Gathering Work Vs Status (Percentages of Respondents Checked)

Reasons for nondelegation	Status								Total	r_s
	A B C & D	E	F	G	H	I	J	K L M & N		
A No assistants	40.0	35.0	47.2	33.3	19.2	48.8	43.9	54.4	42.0	-0.4762
B Difficulty in communicating needs	40.0	5.0	5.6	10.0	11.5	2.4	12.2	12.5	9.5	-0.0714
C Do not believe in others collecting information	40.0	5.0	0.0	15.0	7.7	9.8	9.8	10.7	9.8	-0.0536
D Difficult to train others	20.0	20.0	19.4	11.7	15.4	9.8	4.9	5.4	11.2	0.9464*
E Do not want to disturb others	60.0	5.0	22.2	15.0	23.1	26.8	19.5	14.3	18.9	0.2143
F Searching process is as important as information itself	60.0	60.0	44.4	58.3	57.7	31.7	29.3	12.5	39.6	0.9226*

Key : *, The correlation is statistically significant at $p = 0.05$ as per a two tailed t-test with $df = 6$.

Note : Status A to N are in the decreasing order of the hierarchical rank.

TABLE 6.36 : Reasons of Non-delegation of Information-Gathering Work Vs Qualifications
(Percentages of respondents checked)

Reasons for Non-delegation	Qualifications						
	U	D	B	M	P	Total	x^2
A No assistants	58.1	37.0	35.0	38.2	46.2	41.8	9.2842
B Difficulty in communicating needs	8.2	13.0	11.3	6.6	0.0	9.2	2.3476
C Do not believe in others collecting information	11.5	7.4	8.8	11.8	15.4	10.2	1.3556
D Difficult to train others	4.9	9.3	17.5	7.9	23.1	10.9	8.6661
E Do not want to disturb others	14.8	25.9	21.3	15.8	30.8	19.7	4.1266
F Searching process is as important as information itself	11.5	35.2	47.5	56.6	61.5	40.5	34.1282*

Key : * The association is statistically significant at $p = 0.05$ as the tabulated value of x^2 at $p = 0.05$ for 4 degrees of freedom is 9.488. *Qualifications*-U, Undergraduate; D, Diploma holder; B, Graduate; M, Postgraduate; P, Doctorate.

Note : The contingency tables on which chi-square test is made are not shown.

TABLE 6.37 : Reasons for Nondelegation of Information-Gathering Work Vs Nature of Work

(Percentages of Respondents Checked)

Reasons for Nondelegation	Nature of Work							Total	x^2
	A	B	C	P	D	E	F		
A No assistants	31.6	60.6	31.5	45.5	51.3	36.0	38.5	42.2	12.8711*
B Difficulty in communicating needs	15.8	12.1	6.7	9.1	15.0	4.0	0.0	9.7	5.9301
C Do not believe in others collecting information	10.5	9.1	9.0	18.2	7.5	8.0	16.7	9.3	2.2053
D Difficulty to train others	21.1	9.1	15.7	0.0	7.5	16.0	0.0	11.5	8.1854
E Do not want to disturb others	21.1	21.2	14.6	5.5	16.3	32.0	16.7	19.7	13.0336*
F Searching process is as important as Information itself	63.2	42.4	51.7	36.4	21.3	36.0	39.8	39.8	21.3846*

Key : * The association is statistically significant as the tabulated value of x^2 at $p = 0.05$ for $df = 6$ is 12.592; *Nature of work*- A, Management/Supervision; B, Planning/System analysis; C, Design and Development; P, Those checked both C and D; D, Fabrication and testing; E, Operational activity; F, Others.

Note The contingency tables on which chi-square test is made are not shown.

TABLE 6.38 : Reasons for Nondelegtion of Information-Gathering Wrok Vs Specialisation
(Percentages of Respondents Checked)

Reasons for Non-delegation	Specialisation							
	A	C	E	F	H	I	Total	x^2
A No assistants	33.3	50.0	44.9	50.0	57.1	38.2	41.4	2.8512
B Difficulty in communicating needs	4.8	0.0	7.7	8.3	14.3	13.0	10.0	3.6419
C Do not believe in others collecting information	19.1	10.0	11.5	8.3	0.0	11.4	11.6	2.2174
D Difficult to train others	23.8	0.0	7.7	16.7	14.3	12.2	11.6	5.9388
E Do not want to disturb others	33.3	30.0	19.2	16.7	28.6	17.1	19.9	4.0616
F Searching processis as an important as information itself	52.4	50.0	30.8	66.7	28.6	38.2	38.7	8.5386

Key : *Specialisation* - A, Physics; C, Mathematics; E, Mechanical Engg.; F, Aeronautical and Structural Engg.; H, Electrical Engg.; I, Electronics.

Note : 1. There is no statistically significant association as the tabulated value of x^2 at $p = 0.05$ for $df = 5$ is 11.670.

2. The contingency tables on which the chi-square test is made are not shown.

TABLE 6.39 : Reasons for Non-delegation of Information-Gathering Wrok Vs Experience

Reasons for Non-delegation	Average experience of those		
	Checked	Not checked	x^2
A No assistants	9.10	8.30	7.6396
B Difficulty in communicating needs	10.33	8.46	8.0560
C Do not believe in others collecting information	7.78	8.74	5.1727
D Difficult to train others	8.24	8.69	2.6685
E Do not want to disturb others	8.82	8.58	2.0080
F Searching process is as important as information itself	8.51	8.74	7.4715

Notes : 1. None of othe associations were statistically significant at 0.05 significance level for 5 degrees of freedom (df) in a two tailed test as the tabulated critical value of x^2 is 11.670.

2. The contingency tables on which the chi-square test is made are not shown.

TABLE 6.40 : Reasons for Non-delegation of Information- Gathering Work Vs Professional Activities and Achievements Index

Reasons for Non-delegation	Average Professional Activities and Achievements Index Value of those		
	Checked	Not checked	x^2
A No assistants	5.52	8.12	5.5267
B Difficulty in communicating needs	7.32	7.14	10.7256
C Do not believe in others collecting information	6.91	7.18	2.2075
D Difficult to train others	8.87	6.85	5.5938
E Do not want to disturb others	8.54	6.75	5.0861
F Searching process is as important as information itself	9.77	4.86	11.8895*

Key : * The association is statistically significant at 0.05 siginificance level for 5 degrees of freedom *(df)* in a two tailed test as the tabulated critical value of X^2 is 11.670.

Note : The contingency tables on which the chi-square test is made are not shown.

few had occasional or frequent accidental acquisition of information. However, all the interviewees felt that it is a very important way of acquiring information. There is no indication of any difference in accidental acquisition of information by the IST from that of others (Rosenbloom, *et al.* 1965; Columbia University, 1958; Scott, 1959, p 114; Scott, 1960, p 29; Bernard et al. 1963/1964; Shuchman, 1981, p 15). Respondents have narrated specific incidents of accidental (or chance or unplanned) acquisition of information while skimming through journals, unintentionally looking here and there, browsing library shelves and reorganising personal and departmental collections. To quote one, "Systematic search for information has not been of use... I have found the library very good for browsing. Browsing has indeed been very valuable in view of interdisciplinary nature of my work". Yet another stated that, "Sometimes, skimming through the current journals on display springs surprises by a pleasant exposure to the information sought".

The discussion with the respondents on how they browse in unrelated, loosely related, interdisciplinary and core areas of their interest and concern, (though, the respondents perceived these terms slightly differently), showed that they do not regularly browse in unrelated and loosely related areas except when specific references are found or when cover page or some other feature of a journal or book catches their attention. However, some of them regularly browsed documents of interdisciplinary areas but majority browsed only in the core areas.

6.5 Time Spent on Information-gathering Activities

The amount of time normally spent by the IST in gathering work-related information, literature and references both from formal as well as informal sources including time spent in literature search, retrieval of references, locating documents, purposive interpersonal communication has been ascertained through the questionnaire (Q. No.5). The responses presented in Table 6.41 show that a largest number (37.9 per cent) of the space technologists spend 7 to 10 hours per week. If the range is considered from 4 to 15 hours per week, nearly 80 per cent of the respondents fall under this group. In other words, (taking 40 hours per week as the normal working time) the IST spend about one-fourth of their working time in seeking work related information, which is close to that found by Raitt (1984, p. 151-154) and slightly lower than the time spent by the respondents of Schussel (1969), Halbert and Ackoff (1959), Jahoda (1969) and Gilmore et. al. (1967, p. 41-42). The mean time of 9.24 hours

TABLE 6.41 : Time spent on Information-Gathering Activities

	Time	No.	%
A	More than 20 hours per week	22	4.2
B	16–20 hours per week	28	5.3
C	11–15 hours per wek	118	22.3
D	7–10 hours per week	201	37.9
E	4–6 hours per week	104	19.6
F	Less than 4 hours per week	57	10.7
	Total	530	100.0

Notes : 1. Three not answered and two invalid responses have been excluded.

2. The number of anonymous responses to the categories A to F respectively are 4, 1, 7, 6, 7 and 5.

3. Taking 21, 18, 13, 8.5, 5 and 3 as respective medians of the ranges A to F, the mean worked out to 9.24 hours per week and the standard deviation to 4.58.

TABLE 6.42 : Relative Time Spent on Informal Sources of Information Out of Total Time Spent on Information-Gathering Activities

Percentages of Time Spent on Informal and Interpersonal Sources	No.	%	Cumulative %
00	7	2.3	2.3
01–10	43	14.0	16.3
11–20	41	13.3	29.6
21–30	74	24.0	53.6
31–40	44	14.3	67.9
41–50	57	18.5	86.4
51–60	22	7.1	93.5
61–70	10	3.2	96.7
71–80	9	2.9	99.6
81 90	0	0.0	99.6
91 100	1	0.3	99.9
Total	308	99.9	99.9

Notes : 1. Two hundred and twenty not answered and seven invalid responses are not shown.

2 The number of anonymous responses belonged to the above ranges respectively are 0, 2, 2, 4, 3, 5, 1, 0, 1, 0.and 0.

per week spent on information-gathering activities is higher than that (7 hours per week) found in another Indian study (Garg and Ashok Kumar, 1984, p. 71). The standard deviation is 4.58 hours per week.

The percentage of time spent on informal and interpersonal sources of information (for work related information) as compared to formal and documentary sources is shown in Table 6.42. It is clear from the table that only 13.5 per cent of the respondents spent more than 50 per cent of the information-gathering time on informal sources. On the other hand, 86.4 per cent spent more than 50 per cent of their time on formal sources which conforms to the recent finding of Raitt (1984, p. 153) and differs from that found by Halbert and Ackoff (1959). It is also significant that nearly 30 per cent of the respondents spent less than 20 per cent of their time on informal sources, but only one person (0.3 per cent) spent less than 20 per cent of his time on formal sources.

By and large, the IST have spent slightly more time on formal and documentary sources of information than informal and interpersonal sources. Thus, if one has to judge by the amount of time spent, it is found that the IST have depended slightly more on formal and documentary sources of information than on informal and interpersonal sources of information. The space technologists are unable to spend as much time as they wished due to pressure of their routine works and activites. This observation is very much reflected in more than one-fourth of weekly summary sheets returned by the respondents and in the free flow discussion the investigator had with them. Raitt (1984, p. 228) also found that 69 per cent of the respondents did not spend as much time as they would like to. A typical respondent, in the weekly summary sheets, writes for three of the six weeks as follows : " I could not find enough time for information-activites as I was more involved in routine work such as development and testing of . . ." Another spells out in the open-ended question of the questionnaire that "it is not possible to have information-gathering activity on a regular basis due to pressure of work. Hence, it is done as and when time permits. The time spent on information-gathering activity answered at Q.No.5.1 is only an average over a year". Yet another says "pressure of routine works prevents me from collecting as much as possible and as regularly as possible I would like to do".

6.6 Correlation of Time Spent on Information-gathering Activities with (Selected) User-characteristics

Table 6.43 presents the results of correlation/association tests of the amount of time spent on information-gathering activities with user-characteristics. The contingency tables of time spent on information-gathering activities with status, qualifications, nature of work, and specialisation are shown in Tables 6.44, 6.45, 6.46 and 6.47 respectively. There exists a significant relation of all the six user-characteristics[10] (except nature of work) with the amount of time spent on information-gathering activities, which is quite contrary to findings of an earlier Indian study (Garg and Ashok Kumar, 1984, p.71).

It is clear from Table 6.46, that the managers/supervisors do not differ from others in time spent on information-gathering activities though Raitt (1984, p. 228) found (without any statistical significance test) that those in supervisory positions have spent less time on reading than those in non-supervisory positions. The amount of time spent on information-gathering activities has correlated highly and positively (r_s = 0.76) with status (Table 6.44) and related positively and linearly with qualifications (Table 6.45) of the IST. In other words, the higher the status and higher the qualifications, the more time the user is likely to spend on information-gathering activities. This behaviour can be explained as follows : As one becomes more qualified, one tends to become more specialised and information-conscious. Hence, one tries to spend more time on information related activities. At the same time, as one moves up in the hierarchy, one's professional recognition increases and one spends more time on information-gathering activities to maintain the professional competence. It is also true that qualifications and status of the users are higly related.

Further, physicists and mathematicians (or science personnel) have spent significantly more time on information-gathering activities compared to engineers (Table 6.47) excepting electronics engineers. This is in conformity with the finding of Raitt (1985, p. 322;1984, p. 228) that "... engineers spent less time than scientists ...". It has almost been axiomatic that scientists spend more time in gathering information than engineers and technologists. Further, the length of experience correlated negligibly, but negatively ($r = -0.10$) with the amount of time spent on information-gathering activity contradicting Raitt's (1985, p. 322, 1984, p. 228) finding that "... older staff spent more time in searching for relevant information ...". The professional activities and achievements correlated definitely and positively ($r = 0.36$) with the amount of time spent on information-gathering activities. Though, Hall (1972, p. 158 and 201) found a very weak relation between any performance measure or technical discussion frequency ('discussion stardom') and the amount of

TABLE 6.43 : Results of Association/Correlation Tests of Time Spent on Information-Gathering Activities with User-characteristics

User-characteristics	Degrees of Freedom	Tabulated value of x^2 at $p = 0.05$	$r_s/x^2/r$
Status *(r_s)*	6	—	0.76*
Qualifications *(x^2)*	12	21.03	52.21*
Nature of work *(x^2)*	18	28.87	27.45
Specialisation *(x^2)*	15	24.996	26.75*
Experience *(r)*	394	—	0.10*
Professional Activities and Achievements Index *(r)*	193	—	0.36*

Key : *, The association/correlation is statistically significant at 0.05 significance level.

TABLE 6.44 : Time spent on Information-Gathering Activities Vs Status

	Time spent in hours per week								
Status	**>20**	**16-20**	**11-15**	**1-10**	**4-6**	**<4**	**Total**	**Mean***	**S.D.**
A, B, C, & D	1	1	4	7	4	1	18	9.64	4.61
E	4	4	14	14	11	1	48	10.73	4.99
F	1	6	20	23	14	4	68	9.80	4.34
G	2	9	26	38	21	8	104	9.56	4.42
H	6	2	12	19	12	5	56	9.90	5.26
I	3	2	16	26	9	6	62	9.53	4.40
J	2	2	12	32	11	10	69	8.57	4.11
K, L, M & N	1	1	5	36	19	19	81	6.94	3.44
Total	20	27	109	195	101	54	506	9.18	4.54

($r_S = 0.7619$, $df = 6$, $p < 0.05$)

Key : *, While working out mean 21,18,13,8. 5,5 and 3 hours are taken as medians of respective ranges.

Note : The status A to N are in the descending order of hierarchical rank.

TABLE 6.45 : Time Spent on Information-Gathering Activities Vs Qualifications

Qualifications		Time Spent in hours per week					Mean*	S.D.
		>15	11-15	7-10	<7	Total		
U	Undergraduate	3	7	40	44	94	6.95	3.82
D	Diploma holder	3	25	43	27	98	8.83	3.53
B	Graduate	14	35	55	48	152	9.26	4.63
M	Post-graduate	22	38	56	32	148	10.37	4.64
P	Doctorate	2	11	5	5	23	10.89	4.58
	Total	44	116	199	156	515	9.15	4.47

($x^2 = 52.206$, $df = 12$, $p < 0.05$)

Key : * For working out mean, 21,18,13,8.5, 5 and 3 hours are taken as the respective medians of the six ranges as per questionnaire.

TABLE 6.46 : Time Spent on Information-Gathering Activities Vs Nature of Work

Nature of work		Time spent in hours per week					Mean*	S.D.
		> 15	11-15	7-10	<7	Total		
A	Management/Supervision	5	9	13	15	42	9.27	4.66
B	Planning/System analysis	9	15	22	11	57	10.79	4.89
C	Design and development	14	44	62	39	159	9.67	4.23
P	Both C & D	2	5	15	6	28	9.05	4.16
D	Fabrication and testing	6	24	63	46	139	8.25	3.98
E	Operational activity	2	6	12	15	35	7.83	4.52
F	Others	3	1	8	8	20	8.70	4.95
	Total	41	104	195	140	480	9.14	4.42

($x^2 = 27.4458$, $df = 18$, $p > 0.05$)

Key : *, For working out mean, 21, 18, 13, 8.5, 5 and 3 hours are taken as the respective medians of the six ranges as per questionnaire.

TABLE 6.47 : Time Spent on Information-Gathering Activities Vs Specialisation

Specialisation	Time spent in hours per week					Mean*	S.D
	>15	11-15	7-10	<7	Total		
A Physics	7	10	15	10	42	10.60	5.39
C Mathematics	3	6	7	3	19	11.03	4.43
E Mechanical Engg.	3	23	57	52	135	7.79	3.65
F Aeronautical Structural Engg.	0	6	4	9	19	8.15	3.60
H Electrical Engg.	1	4	7	5	17	8.73	4.09
I Electronics	23	50	88	59	220	9.52	4.54
Total	37	99	178	138	452	9.08	4.44

($x^2 = 26.7464$, $df = 15$, $p < 0.05$)

Key: *, For working out mean, 21, 18, 13, 8.5, 5 and 3 hours are taken as the respective medians of the six ranges as per questionnaire.

TABLE 6.48 : Results of Association/Correlation Tests of Percentages of Information-Gathering Time Spent on Informal and Interpersonal Sources of Information with (Selected) User-characteristics

	Status (x^2)	Qualifications (x^2)	Nature of works (x^2)	Speciali-sation (x^2)	Experience (r)	Professional Activities and Achievements Index (r)
x^2 / r	19.71	8.33	8.08	9.57	–0.12*	–0.07
Degrees of freedom (df)	14	8	12	10	247	125
Tabulated Value of x^2 at $p = 0.05$	23.69	15.51	21.03	18.31	—	—

Key : * The association/correlation is statistically significant at 0.05 significance level.

time spent in reading technical literature, the present study has established a definite positive correlation of amount of time spent on information-gathering activities with professional activities and achievements of the respondents.

The relative time spent on informal and interpersonal sources of information as compared to formal and documentary sources of information has also been tested for correlation/association with user- characteristics and results shown on Table 6.48. The individual contingency tables of percentage of time spent on informal sources of information against status, qualifications, nature of work and specialisation are respectively presented in Tables 6.49, 6.50, 6.51 and 6.52. It is evident from Table 6.48 that the percentage of time spent on informal and interpersonal sources of information is independent of all the six user-characteristics,[11] except a slight negative correlation (r=-0.12) shown in case of length of experience. In other words, the relative time spent on formal and informal sources of information did not vary with the user-characteristics.

6.7 Variation in Time Spent on Information-gathering Activities During Different Stages of Projects

As reflected in summary sheets and responses to interview there is diverse opinion about information-requirement as well as the amount of time spent on information-gathering at different phases/stages of projects or activities. However, some general clues are as follows. In case of TDP, the information-requirement and time spent on information-gathering is considerably more during the initial phase of conception, definition and system study than at later stages. In fact, the need for information as well as time spent in gathering information decreases as the work progresses. In case of large (regular) projects, the same phenomenon holds good except for those who have not involved in the conception and definition of the project or when a project has been given to them by others. A senior project engineer who is involved from the beginning in large projects, has estimated that he spent about 15 hours a week during initial phase of systems study and an average of 3 hours a week during project implementation phase. However, engineers who were not involved in the conception and definition of projects, felt the reverse, i.e., their information requirements as well as the amount of time spent on information gathering is very less in the initial stages of project execution and the same increases gradually, but not necessarily till closure of the project.

For a regular on going activity which is not time bound, no variations

TABLE 6.49 : Percentage of Time Spent on Informal and Interpersonal Sources of Information Vs Status

Status	Percentage of Time Spent			Total
	<26	26-75	>75	
A B C & D	9	8	1	18
E	21	21	0	42
F	18	26	1	45
G	34	41	1	76
H	12	22	1	35
I	19	20	0	39
J	4	23	0	27
K L M & N	5	14	0	19
Total	122	175	4	301

(x^2=19.7051, df=14, p>0.05)

Note : The status A to N are in the descending order of hierarchical rank.

TABLE 6.50 : Percentage of Time Spent on Informal and Interpersonal Sources of Information Vs Qualifications

	Qualifications	Percentage of Time Spent			Total
		<26	26-75	>75	
U	Undergraduate	7	12	1	20
D	Diploma holder	15	35	0	50
B	Graduate	46	60	1	107
M	Postgraduate	44	60	3	107
P	Doctorate	10	8	0	18
	Total	122	175	5	302

(x^2 = 8.3268, df = 8, p > 0.05)

TABLE 6.51 : Percentage of Time Spent on Informal and Interpersonal Sources of Information Vs Nature of Work

Nature of work		Percentage of Time Spent			Total
		<26	26-75	>75	
A	Management/Supervision	13	21	1	35
B	Planning/System analysis	21	21	0	42
C	Design and Development	41	62	1	104
P	Both C and D	8	9	1	18
D	Fabrication and testing	21	33	0	54
E	Operational activity	6	11	0	17
F	Others	5	6	0	11
	Total	115	163	3	281

($x^2 = 8.0788$, $df = 12$, $p > 0.05$)

TABLE 6.52 : Percentage of Time Spent on Informal and Interpersonal Sources of Information Vs Specialisation

Specialisation		Percentage of Time Spent			Total
		<26	26-75	>75	
A	Physics	13	18	0	31
C	Mathematics	1	9	0	10
E	Mechanical Engineering	29	33	2	64
F	Aeronautical and Structural Engineering	7	6	0	13
H	Electrical Engineering	5	4	0	9
I	Electronics	55	81	1	137
	Total	110	151	3	264

($x^2 = 9.5718$, $df = 10$, $p > 0.05$)

Note :By grouping physicists and mathematicians as pure science personnel and the rest as engineering personel the following contingency table is obtained:

Specialisation	Percentage of Time Spent			Total
	<26	26-75	>75	
Science	14	27	0	41
Engineering	96	124	3	223
Total	110	151	3	264

(x^2=1.8828, df=3, p>0.05)

in time spent on information gathering is noticed. For a time bound adhoc activity like writing a paper or report, more time for informaation collection is spent as one nears the deadline.

For an activity in the alien field there appears to be need for spending more time in seeking information in the begining. On the other hand, in case of an activity in one's own field it is either more or less uniform or slightly more towards the end. Similarly for a new design, maximum information is needed in the initial stages and hence more time is spent in seeking information in such a project which requires modification of an existing system.

During conception of a project the more time spent on information gathering, implies more time for discussion than literature search. However, as the detailed definition and study starts, the necessity of discussion may become less than that of the literature. Above all, the respondents have clearly underlined that the phases or stages of a project or problem solving activity need not be necessarily sequential or linear and independent.

NOTES

1. The 'Other means' used in seeking bibliographic information by the respondents are (the number of respondents is noted in brackets): departmental collection (2), other libraries (1), obtaining conference papers and reprints from authors (2), inter-library loan facility (1), literature search by juniors (2), acquiring trade catalogues from manufacturers (2) and accidental acquisition (1).
2. Consulting library staff for references by women respondents did not differ from that of men respondents (Sridhar, 1987).
3. Consulting colleagues for references is also independent of sex of the respondents (Sridhar, 1987).
4. Even in consulting experts in the field for references much difference among men and women space technologists is not noticed (Sridhar, 1987).
5. Women space technologists do not differ significantly from men space technologists in degree of delegation of information-gathering work (Sridhar, 1987).
6. Twelve respondents who gave 'other reasons', included those close to category D in Table 6.17, and miscellaneous reasons like 'easy access to other libraries', 'nature of work' 'to get references/information quickly *i.e.*, speed', 'while working in more than one project/assignment, delegated approach gives the best results', 'lack of patience towards formal sources', 'when work is more familiar to others', etc.
7. The degree of delegation of factual, design and physical data collection is significantly and positively related to the degree of need for such data ($x^2 = 19.51$, $df = 4$, $p < 0.001$).
8. The degree of delegation of retrospective and exhaustive search for references is significantly associated with the degree of requirement of state-of-the-art and review literature ($x^2 = 9.66$, $df = 4$, $p < 0.05$).

9. The additional reasons for non-delegation of information-gathering work included expressions close to serendipity value of information-searching process (7) and not having need to delegate (7) as well as other difficulties (4).
10. Women space technologists do not differ from men space technologists in amount of time spent on information-gathering activities (Sridhar, 1987).
11. The relative time spent on informal and interpersonal sources of information compared to formal and documentary sources of information has no significant difference between the sexes (Sridhar, 1987).

7

DEPENDENCE ON SOURCES OF INFORMATION BY THE INDIAN SPACE TECHNOLOGISTS

7.1 Dependence on Formal and Documentary Sources of Information

The extent to which the Indian space technologists depend on various formal and documentary sources of information is reflected in the data noted in Table 7.1 as gathered in response to the questionnaire (Q.No.3.1). It is clear from the Table that theses and dissertations (D), followed by conference proceedings and papers (C), and standard and patent specifications (G), respectively are not useful sources of information for 31 per cent, 20.4 per cent and 19.9 per cent of the respondents. In addition almost an equal number of the respondents (*i.e.*, 31.8 per cent, 18.3 per cent and 27.2 per cent respectively) have felt that they rarely depend on these sources. Further, 27.4 per cent of respondents depend rarely and another 17.4 per cent do not depend on reprints and preprints from fellow professionals (I). In terms of percentage of the users depending (irrespective of degree of dependence) as against not depending on the formal sources, books (A and B) top the list followed by journals (E), reports (F), personal collection (K), trade literature (J) and official documents (H). However, if one considers the degree of dependence, a large number of respondents (156) indicate their high dependence (Code 4) on periodicals/journals (E) followed by books including handbooks and reference books (A and B), trade literature/product catalogues from suppliers and manufacturers (J), technical/RandD reports (F) and personal collection (K). The Fig. 7.1 pictorially depicts the trends of dependence on formal sources of information. The ranking of formal sources of information by the IST in

TABLE 7.1: Dependence on Formal and Documentary Sources of Information

Formal and Documentary Sources of Information	Non-Dependence (0)		Degree of Dependence				Total		Mean	S.D.
	No.	%	1	2	3	4	No.	%		
A. Books (other than B)	11	2.1	37	152	223	104	516	97.9	2.71	0.92
B. Hand-books/reference books	11	2.1	42	187	180	105	514	97.9	2.62	0.96
C. Confernece proceedings and papers	104	20.4	94	122	139	55	410	79.6	1.90	1.28
D. Theses and dissertations	158	31.0	162	128	51	10	351	69.0	1.20	1.05
E. Periodicals/journals	14	2.7	42	134	177	156	509	97.3	2.80	1.04
F. Technical/ R & D reports	29	5.5	50	132	223	89	494	94.5	2.56	1.06
G. Standard and patent specifications	101	19.9	138	130	86	53	407	80.1	1.71	1.25
H. Offical documents including internal reports of ISRO and departmental collection	49	9.5	106	195	124	40	465	90.5	2.00	1.07
I. Reprints and preprints from fellow professionals	88	17.4	139	143	83	54	419	82.6	1.76	1.22
J. Trade literature/product catalogues from suppliers and manufacturers	41	7.8	104	128	150	97	482	92.2	2.29	1.22
K. Personal collection	35	6.8	110	151	125	94	480	93.2	2.26	1.17
L. Other formal sources	170	89.9	9	1	7	2	19	10.1	0.21	0.72

Key : 0, No dependence; 1, Rare dependence; 2, Occasional dependence; 3, Frequent dependence; 4, High dependence; S.D.,Standard deviation.

Notes : 1. Three hundred and thirty-two respondents have kept open-ended category 'Other formal sourcers' (L) blank.

2. A maximum of 27 have not answered and upto one made invalid response to any category and they are excluded.

3. The mean of anonymous reponses to categories A to L respectively are 2.77, 2.77, 1.41, 0.97, 2.20, 2.43, 1.67, 1.67, 2.34, 2.34, 2.50 and 0.23.

terms of their dependence as per weighted mean is as follows:

Rank	Code	Formal and Documentary Sources of Information
1	E	Journals [1]
2	A	Books
3	B	Handbooks/reference books
4	F	Reports
5	J	Trade Literature
6	K	Personal Collection
7	H	Official Documents[2]
8	C	Conference Proceedings
9	I	Reprints and Preprints[3]
10	G	Standard and Patent Specifications[4]
11	D	Theses and Dissertations

By adding/subtracting one standard deviation from/to the mean, we can statistically say that about 68 per cent of the respondents have shown dependence on journals, books, reference books, reports, trade literature and personal collection in the range of 1.76-3.84, 1.79-3.63, 1.66-3.58, 1.5-3.62, 1.07-3.51 and 1.09-3.43 respectively on a five point scale.

The above ranking places formal sources such as journals, books, trade catalogues, reports, personal collection and official documents as significant sources and the rest as less significant sources of information for the IST. The rank order of dependence on formal sources of information is close to that found by Tornudd (1959), Herner and Myatt (1954) and Davis (1965, p.30-34). However, Raitt (1984, p.253) found higher rank for reports than books. Books have occupied higher rank than reports in the present study presumably due to large number of lower-level space technicians covered. Interestingly, in congruence with the high ranked purpose of seeking information for keeping abreast with the latest developments and the high ranked need for S and T news and basic S and T information, the dependence on journals as a formal source of information has naturally ranked top.

'Ease of use' has been an important criterion in accessing a source of information by the IST as reflected in typical statements like "my information-gathering habits are mostly dependent on things easily come across in everyday life". A large number of the respondents have described and discussed the significance of trade literature/catalogues/ product information during interviews, in summary sheets and open

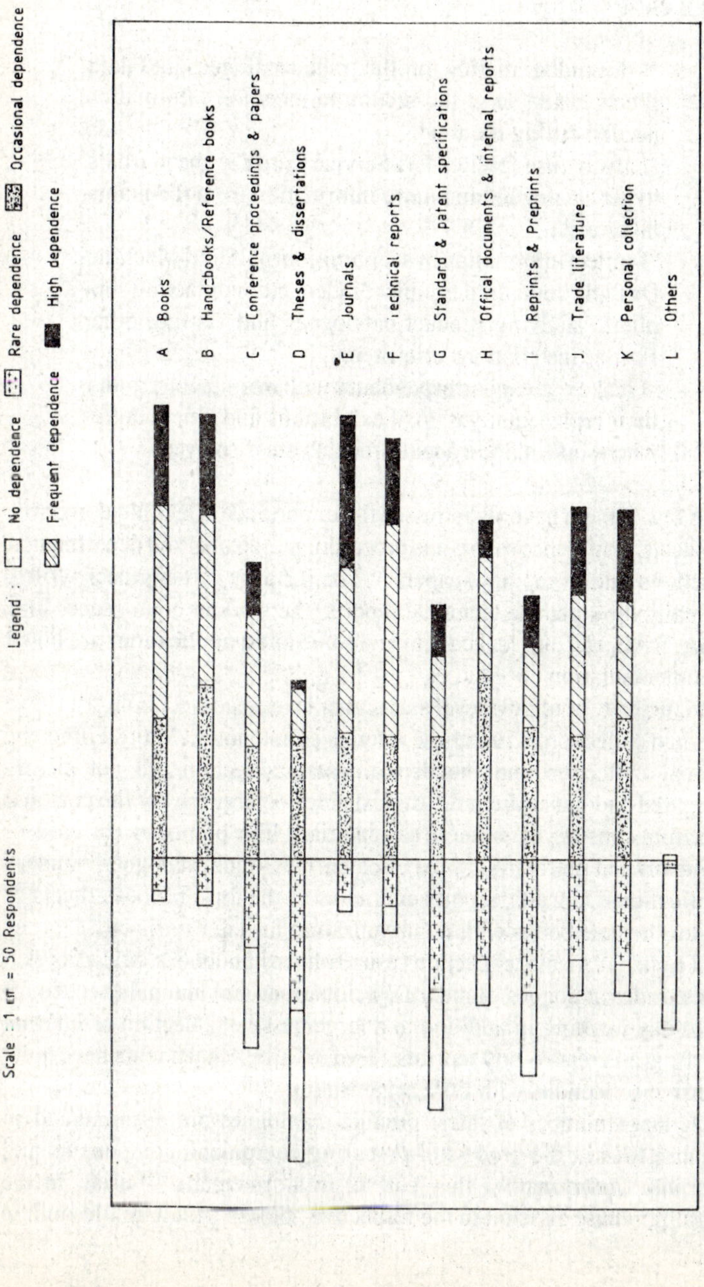

Fig. 7.1 : Dependence on Formal and Documentary Sources of Information

ended question of the questionnaire. The following statements of the users make it clear:

> "I depended mainly on the trade catalogues and data sheets available in the section to meet the information needed during the week."
> "I always use the 'Readers Service Cards' in the journals to get catalogues and more information from the manufacturers".
> "I collect information by scanning journals like 'Machine Design', 'Industrial Product Finder', etc., contacting suppliers, studying product catalogues and visiting other similar installations elsewhere."
> "I make regular correspondence with manufacturers and their representatives, visit exhibitions and other centres where information about products are displayed."

A few of them have also stressed the importance of technical reports, data sheets, conference papers and proceedings, personal and departmental collections and even newspapers. For instance, an engineer writes, "normally through the technical reports, the work at other places like NASA, RAE, etc., are looked into and some latest applications are noted for implementation".

A majority of middle level and higher level space technologists have their own collection of literature, references and notes. Many a time the personal collection and the departmental collection are not clearly segregated and hence are indistinguishable. A large part of the personal collection consists of either free materials like preprints (in case of Astronomy and Astrophysics area), free journals, trade catalogues, standard specifications, reprints, reports and notes in the form of loose sheets or registers or note books with no organisation for later retrieval.

The departmental collections usually have handbooks, priced as well as free trade catalogues, standard specifications and manuals needed for day-to-day working in addition to a large passive collection of internal reports, gratis reports and reprints. Two or three departments have quite extensive collections with little organisation.

A large number of latest product catalogues are accumulated in purchase-files in the process of procuring equipment, components and materials. Unfortunately, they end up in a systematic 'burial' in the central purchase division in the respective 'closed' purchase files unless the indentor takes extra care to remove and transfer them to an appropriate collection.

The significance of personal and departmental collections to the IST is reflected in the following statements extracted from the weekly summary sheets and responses to open-ended question in the questionnaire:

> "I get most of the information required for my work from the departmental collection and from books in the library."
> "For my work of co-ordination and interface control, first I will look into the internal design documents and if the information is not sufficient, I will contact the designers."
> "Almost all information required by us are in the MIL specifications kept in the division."
> "I get technical information from catalogues supplied to me by the manufacturers."
> "The information required for my work is gathered mostly from MIL specifications, books, trade catalogues received from the manufacturers and from my senior colleagues."
> "I get all the information required from the MIL specifications and some information regarding components, etc., from manufacturers catalogues."
> "I try to maintain my own reprints, books and journals library in the division with a proper cataloguing."

7.2 Correlation of Dependence on Formal and Documentary Sources of Information with User-characteristics

The overall results of association/correlation tests of dependence on formal and documentary sources of information with user-characteristics are shown in Table 7.2. Tables 7.3, 7.4, 7.5, and 7.6 respectively present the weighted mean dependence on formal and documentary sources of information against status, qualifications, nature of work and specialisation of the IST. As could be seen from Table 7.3 the dependence of the space technologists on conference proceedings and papers ($r_s = 0.98$), theses and dissertations ($r_s = 0.95$), technical reports ($r_s = 0.85$), official documents and internal reports ($r_s = 0.67$), reprints and preprints ($r_s = 0.96$) and personal collection ($r_s = 0.84$) increases linearly as the status of the user increases but dependence on standard and patent specifications ($r_s = -0.81$) decreases linearly as the status of the user increases.

TABLE 7.2 : Association/Correlation Test of Dependence on Formal and Documentary Sources of Information with (Selected) User-characteristics

	Formal and Documentary Sources of Information	Status (r_s) $df = 8$	Qualification (x^2) $df = 16$	Nature of work (x^2) $df = 24$	Specialisation (x^2) $df = 20$	Experience (r) $df = 390$	Professional Activities and Achievements Index (r) $df = 193$
A	Books	–0.25	23.76	40.31*	15.32	–0.11*	-0.08
B	Handbooks/reference books	0.59	54.00*	29.81	35.37*	–0.07	0.04
C	Conference Proceeding and Papers	0.98*	190.01*	90.64*	67.83*	0.11*	0.44*
D	Theses and dissertations	0.95*	82.37*	61.20*	34.53*	–0.02	0.19*
E	Journals	0.55	64.75*	32.49	25.28	0.15*	0.21*
F	Technical reports	0.85*	38.61*	25.61	28.37	0.09	0.10
G	Standard and patent specifications	–0.81*	89.96*	49.81*	62.41*	0.11*	–0.21*
H	Official documents and Internal reports	0.67*	72.23*	40.07*	33.89*	–0.11*	0.09
I	Reprints and Preprints	0.96*	100.63*	56.85*	52.88*	0.13*	0.37*
J	Trade literaturte	–0.28	18.86	48.34*	40.26*	0.10	–0.12*
K	Personal collection	0.84*	86.13*	49.73*	33.07*	0.04	0.23*

Key : *, The association/correlation is statistically significant at 0.05 significance level *(p)* for a two tailed test.

Note : The tabulated values of x^2 at 0.05 significance level *(p)* for a two tailed test for 16, 24 and 20 degrees of freedom *(df)* are 26.296, 36.415 and 31.410 respectively.

TABLE 7.3 : Status Vs Dependence on Formal and Documentary Sources of Information

(Weighted Mean Score)

Formal and Documentary Sources of Information	Status										
	A,B, C & D	E	F	G	H	I	J	K	L	M & N	Total	r_s
A Books	2.60	2.63	2.75	2.82	2.78	2.80	2.67	2.52	2.73	2.58	2.72	-0.2485
B Hand books/reference books	2.38	2.88	2.75	2.79	2.72	2.76	2.38	2.16	2.53	1.80	2.64	0.5909
C Conference proceedings and papers	2.86	2.98	2.83	2.20	1.71	1.33	1.23	0.68	0.68	0.47	1.82	0.9848*
D Theses and dissertations	1.40	1.37	1.52	1.29	1.35	1.09	0.97	0.59	0.50	0.47	1.16	0.9515*
E Journals	3.30	3.29	3.25	2.87	2.63	2.62	2.49	2.08	2.35	2.17	2.82	0.5455
F Technical reports	2.95	3.04	3.03	2.59	2.44	2.37	2.12	2.13	2.28	2.28	2.54	0.8515*
G Standard and patent specifications	1.19	1.86	1.43	1.30	1.55	2.17	2.11	1.88	2.55	2.13	1.72	-0.8061*
H Official documents and internal reports	1.85	2.00	2.37	2.22	2.17	1.95	1.89	1.59	1.59	0.81	2.00	0.6697*
I Reprints and preprints	2.35	2.40	2.32	1.65	1.70	1.46	1.38	1.00	1.18	0.63	1.70	0.9636*
J Trade literature	2.35	2.40	2.44	2.23	2.08	2.22	2.49	2.13	2.66	2.44	2.33	-0.2818
K Personal collection	2.45	2.72	2.70	2.34	2.39	1.86	1.45	1.96	2.02	2.78	2.17	0.8394*

Key : r_s, Spearman rank order correlation; *, the critical value of t at 0.05 significance level in a two tailed t-test for 9 degrees of freedom is 2.262 and hence correlations maked * are found valid at 0.05 significance level.

Note : Status A to N are in the descending order of hierarchical rank.

TABLE 7.4 : Qualifications Dependence on Formal and Documentary Sources of Information ***(Weighted mean score)***

Formal and Documentary Sources of Information	Qualifications						
	U	D	B	M	P	Total	x^2
A Books	2.57	2.90	2.77	2.68	2.91	2.74	23.7648
B Handbooks/ reference books	2.12	2.76	2.87	2.66	2.69	2.65	54.0040*
C Conference proceeding and papers	0.57	1.24	2.05	2.46	3.04	1.82	190.0062*
D Theses and dissertations	0.42	1.06	1.19	1.28	1.65	1.08	82.3715*
E Journals	2.26	2.35	2.79	2.95	3.65	2.70	64.7498*
F Technical reports	2.13	2.33	2.65	2.63	3.30	2.67	38.6106*
G Standard and patent specifications	2.38	2.37	1.56	1.26	1.35	1.76	89.9643*
H Official documents and internal reports	1.27	2.00	2.18	2.24	1.96	2.00	72.2282*
I Reprints and preprints	0.90	1.38	1.66	1.97	2.69	1.62	100.6278*
J Trade literature	2.60	2.55	2.42	2.20	1.91	2.39	18.8624*
K Personal collection	1.69	1.81	2.34	2.39	3.00	2.17	86.1294*

Key : *, The association is statistically significant at $p = 0.05$ as the x^2 value at $p = 0.05$ for 16 degrees of freedom is 26.296; *Qualifications*: U-Undergraduate, D-Diploma holder, B-Graduate, M-Postgraduate, P-Doctorate.

Exactly the same trend of linear positive relation of dependence on conference proceedings, technical reports, internal reports, reprints and preprints and personal collection and linear negative relation of dependence on standard and patent specifications can be seen with the level of qualifications of the IST in Table 7.4. The finding of Shuchman (1981, p. 36) that dependence on internal technical reports is the greatest for those with least education is diagramatically opposite to what is found in the present study. In addition, dependence on reference books and journals are also linearly and positively related to the qualifications of the users.

Table 7.5 reveals that the dependence on books, conference proceedings and papers, theses and dissertations, standard and patent specifications, official documents and internal reports, reprints and preprints, trade literature and personal collection has significantly varied with the nature of work of the space technologists. The dependence on reprints and preprints and personal collection has almost increased linearly as the management/supervision content of the job increased. Planners/system analysts have depended more on conference proceedings and papers, theses and dissertations and internal reports, design, development, fabrication and testing personnel have depended more on books than others. Both managers/supervisors and fabrication and testing personnel have depended more on trade literature than others. The highest degree of dependence on standard and patent specifications is laid by fabrication and testing personnel.

As per Table 7.6 the dependence of the IST on reference books,

TABLE 7.5 : Nature of Work Vs Dependence on Formal and Documentary Sources of Information

(Weighted Mean Score)

Formal and Documentary Sources of Information	Nature of work							Total	x^2
	A	B	C	P	D	E	F		
A Books	2.54	2.75	2.84	2.85	2.81	2.37	2.85	2.76	40.3100*
B Handbooks/ reference books	2.64	2.66	2.83	2.60	2.52	2.58	2.20	2.65	29.8122
C Conference proceedings and papers	2.04	2.30	2.17	2.11	1.17	1.06	1.58	1.79	90.6435*
D Theses and dissertations	1.12	1.43	1.28	0.85	0.73	0.75	1.00	1.06	61.2014*
E Journals	2.88	2.91	2.82	2.57	2.38	2.70	2.60	2.68	32.4876
F Technical reports	2.54	2.63	2.66	2.43	2.44	2.06	2.37	2.52	25.6077
G Standard and patent specifications	1.81	1.24	1.57	1.46	2.26	2.03	1.52	1.77	49.8132*
H Official documents and internal reports	2.05	2.43	2.07	2.07	1.73	1.82	1.80	1.99	40.0667*
I Reprints and preprints	2.07	2.00	1.72	1.81	1.19	1.27	1.60	1.61	56.8491*
J Trade literature	2.63	1.85	2.32	2.48	2.64	2.29	2.21	2.38	48.3414*
K Personal collection	2.53	2.50	2.30	2.33	1.78	1.88	2.15	2.16	49.7252*

Key: *, The association is statistically significant at $p = 0.05$ as the tabulated value of x^2 at $p = 0.05$ for 24 degrees of freedom is 36.415; *Nature of Work* : A-Management Supervision, B-Planning/System Analysis, C Design and Development, D Fabrication and Testing, E-Operational Activity, F-Others.

conference proceedings and papers, theses and dissertations, standard and patent specifications, internal reports, reprints and preprints, trade literature and personal collection has varied with the specialisation of users. Particularly, higher dependence of aeronautical and structural engineers on conference proceedings, theses, technical reports, reprints and personal collection; mechanical engineers on standard and patent specifications; mathematicians on internal reports; electrical and electronics engineers on reference books and trade literature can be noticed. Thus the Raitt's (1984, p. 234) finding that engineers consulted reference books much less than scientists is not true in the present study. But, Shuchman's (1981, p. 36). finding of aerospace engineers using technical reports more than others is supported in the present study.

TABLE 7.6 : Specialisation Vs Dependence on Formal and Documentary Sources of Information

(Weighted Mean Score)

Formal and Documentary Sources of Information	Specialisation							
	A	C	E	F	H	I	Total	x^2
A Books	2.78	2.85	2.66	3.05	2.71	2.79	2.76	15.3164
B Handbooks/ reference books	2.79	2.80	2.40	2.53	2.88	2.83	2.66	35.5738*
C Conference proceeding and papers	2.22	1.68	1.24	2.79	2.00	1.88	1.79	67.8299*
D Theses and dissertations	1.20	1.06	0.82	1.58	1.13	1.11	1.05	34.5256*
E Journals	2.95	2.70	2.44	3.26	3.00	2.70	2.68	25.2845
F Technical reports	2.71	2.21	2.36	3.11	3.00	2.52	2.52	28.3716
G Standard and patent specifications	1.27	1.06	2.36	1.11	1.76	1.66	1.79	62.4128*
H Official documents and internal reports	2.43	2.70	1.75	1.95	2.06	1.93	1.97	33.8942*
I Reprints and preprints	2.10	1.74	1.35	2.32	2.24	1.53	1.60	52.8790*
J Trade literature	2.31	1.61	2.47	1.53	2.63	2.51	2.40	40.2635*
K Personal collection	2.39	2.39	1.98	2.53	2.24	2.16	2.15	33.0683*

Key : *, Association is statistically significant as tabulated value of x^2 at $df = 20, p = 0.05$ is 31.410. *Specialisation* : A-Physics, C-Mathematics, E-Mechanical Engineering, F-Aeronautical and Structural Engineering, H-Electrical Engineering, I-Electronics.

The experience of the space technologists shows (Table 7.2) a slight positive correlation with their dependence on conference proceedings and papers ($r = 0.11$), journals ($r = 0.15$), standard and patent specifications ($r = 0.11$) and reprints and preprints ($r = 0.13$) and slight negative correlation with their dependence on books ($r = -0.11$) and internal reports ($r = -0.11$). On the other hand, professional activities and achievements of the space technologists have shown a moderate positive correlation with dependence on conference proceedings and papers ($r = 0.44$) and reprints and preprints ($r = 0.37$), a low positive correlation with dependence on theses and dissertations ($r = 0.19$), journals ($r = 0.21$) and personal collection ($r = 0.24$), a low negative correlation ($r = -0.21$) with standard and patent specifications and slight negative correlation ($r = -0.12$) with trade literature.

As observed in Chapter 4, 'designing and development of products and components' (Q. No.2 B), 'setting up and use of equipment' (Q. No. 2 C) as purposes of seeking information are moderately intercorrelated ($r = 0.40$).It is expected that 'methods, processes and procedures information' (Q. No. 1 D) and 'product, material, equipment and apparatus information' (Q. No. 1 E) which correlate moderately between themselves ($r = 0.46$) are the type of information required for the above purposes. Both these types of information are expected mostly from 'trade literature' (Q. No.3. 1J) which rank fifth among the formal sources of information. Interestingly, all the intercorrelations among these five factors shown in Appendix-7 are positive and statistically significant except correlation between need for methods, processes and procedures information (Q. No. 1 D) and the degree of dependence on trade literature (Q. NO. 3.1 J). Thus, the information requirements for these purposes are closely related to trade literature and hence depending on changes in the purposes of seeking information and the requirement, dependence on trade literature is likely to vary.

7.3 Dependence on Informal and Interpersonal Sources of Information

The degree of dependence of the space technologists on informal and interpersonal sources of information presented in Table 7.7 based on replies to questionnaire (Q. No. 3.2) shows that as many as 135 or 29.4 per cent of the respondents do not depend on fellow professionals outside ISRO (Q) for information. Also 14.3 per cent of the respondents do not depend on professional meetings, seminars, symposia and lectures (S), as well as educational and training programmes (T) for any information. On the other hand, personal experience (M), superiors (N), peers and colleagues in ISRO (O), subordinates and juniors (P) and results of one's own experiments (R) are the sources on which more than 90 per cent of the respondents depend for information.[5] The rank order of the dependence on informal and inter-personal sources of information based on mean score is as follows:

Rank	Code	Informal and Interpersonal Sources of Information
1	M	Personal experience
2	R	Results of one's/own experiments[6]
3	N	Superiors
4	O	Peers and colleagues in ISRO
5	S	Professional meetings, seminars, symposia and lectures[7]
6	T	Educational and training courses[8]
7	P	Subordinates and juniors
8	Q	Fellow professionals outside ISRO[9]

By adding/subtracting one S.D. to/from mean, we find that about 68 per cent of the respondents have checked personal experience (M), superiors (N), peers and colleagues (O) and results of one's/own experiments (R) in the range of 2.1-4.00, 1.27-3.37, 1.26-2.96 and 1.43-3.73 respectively on a five point scale. The space technologists have depended more on these sources than fellow professionals outside ISRO (Q), subordinates and juniors (P), professional meetings, and seminars (S) and educational and training courses (T).

The intrapersonal sources *i.e.,* personal experience and results of one's/own experiments claim highest dependence by the IST for work-related information. Then come the superiors, peers and colleagues as a second group of highly-depended sources. All other sources have a rare or less than occasional dependence by the IST. Fig. 7.2 depicts through bar charts the pattern of responses about dependence on informal and interpersonal sources of information.

The findings of the present study very much support the recent findings of Shuchman (1982, p. 5) about American engineers that "the primary source of engineering information is largely what the engineer keeps in his head". Leaving the intrapersonal sources (which many earlier studies have not tackled) the most significant informal sources of information for the space technologists are superiors, peers and colleagues

TABLE 7.7 : Dependence on Informal and Interpersonal Sources of Information

Informal and Interpersonal Sources of Information	No Dependence (0)		Degree of Dependence						Mean	S.D.
							Total			
	No.	%	1	2	3	4	No.	%		
M. Personal experience	10	2.0	36	60	163	222	481	98.0	3.12	1.02
N. Superiors	18	3.6	59	199	186	34	478	96.4	2.32	0.91
O. Peers and colleagues in ISRO	22	4.5	70	245	140	15	470	95.5	2.11	0.85
P. Subordinates and juniors	39	8.1	175	172	89	9	445	91.9	1.70	0.92
Q. Fellow professionals outside ISRO	135	29.4	130	117	69	8	324	70.6	1.31	1.10
R. Results of one's/own experiments	24	5.1	67	109	159	116	451	94.9	2.58	1.15
S. Professional meetings, seminars, symposia and lectures	68	14.3	120	152	118	17	407	85.7	1.78	1.08
T. Educational and training courses	68	14.3	125	178	70	33	406	85.7	1.74	1.09
U. Other informal sources	170	91.9	8	4	2	1	15	8.1	0.14	0.54

Key : 0, No dependence; 1, Rare dependence; 2, Occasional dependence; 3, Frequent dependence; 4, High dependence; S.D., Standard deviation.

Notes : 1. Three hundred and forty-two respondents have kept open-ended category 'Other informal sources' (U) blank.

2. A maximum of 75 have not answered and a maximum of 11 made invalid responses to any category and they are excluded.

3. The mean of anonymous responses to categories M to U respectively are 3.14, 2.21, 2.17, 1.76, 1.14, 2.38, 1.41, 1.62 and 0.21.

as found in other studies (Glass and Norwood, 1959; Gralewska-Vickery, 1976, p. 269) Sherwin and Inemson, 1966 and Rosenbloom and Wolek, 1970).

The discussion with respondents emphasised that the "personal experience plays a major role and only when there is a vague resemblance, one is led to formal literature". Further, "due to need for information relating to practical processes in highly specialised areas, the nucleus of information-gathering activities of the space technologists is often found as oneself". "The non-availability of similar information in formal sources may not make much impact on work". Those who depend more on informal sources have claimed to depend mostly on superiors, juniors and colleagues and at times experts outside-the-organisation. Such interpersonal contacts are mainly for opinion, suggestions and references. A doctorate-scientist revealed how the cross fertilisation of ideas he had in the discussions with an agricultural scientist and a geologist (both were his neighbours) triggered his work on developing two equipments he later patented, which have brought him two coveted awards.

7.4 Correlation of Dependence on Informal and Interpersonal Sources of Information with User-characteristics

The results of association/correlation tests for dependence on informal and interpersonal sources of information with the six user-characteristics are presented in Table 7.8. Tables 7.9, 7.10, 7.11 and 7.12 respectively present the mean scores of dependence on informal and interpersonal sources of information against status, qualifications, nature of work and specialisation of the respondents.

As shown in Table 7.9 the status of the space technologists is found to be highly correlated positively with the dependence on personal experience (r_s = 0.88), subordinates and juniors (r_s = 0.71), fellow professionals outside ISRO (r_s = 0.94), results of own experiments (r_s = 0.95), professional meetings and seminars (r_s = 0.98) and negatively with superiors (r_s = –0.95) and educational and training courses (r_s = –0.81) as sources of information. The finding that the space technologists have depended increasingly more on superiors for information as the status decreases supports the findings of Arnet *et al.* (1980, p. 441-448).

Table 7.10 reveals that increased with the level of qualifications, the IST have been increasingly more dependent on personal experience and fellow professionals outside ISRO and decreasingly less dependent on superiors for information in conformity with what Shuchman (1981,

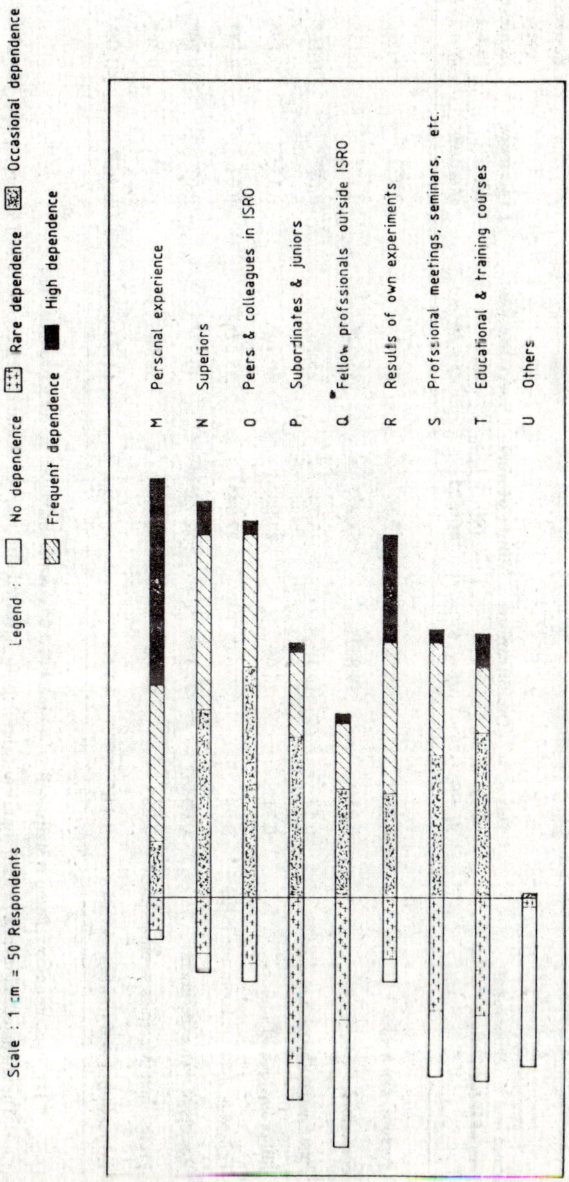

Fig. 7.2 : Dependence on Informal and Interpersonal Sources of Information

TABLE 7.8 : Association/Correlation Test Results of Dependence on Informal and Interpersonal Sources of Information with (Selected) User-characteristics

	Informal and Interpersonal Sources of Information	Status (r_s) $df = 8$	Qualifications (x^2) $df = 16$	Nature of work (x^2) $df = 24$	Specialisation (x^2) $df = 20$	Experience (r) $df = 368$	Professional Activities Achievements Index (r) $df = 182$
M	Personal experience	0.88*	43.56*	27.58	20.33	0.02	0.06
N	Superiors	–0.95*	38.86*	68.97*	35.62*	–0.17*	–0.13*
O	Peers and colleagues in ISRO	0.57	19.76	33.74	25.10	–0.07	0.16*
P	Subordinates and Juniors	0.71*	26.00	24.90	40.70*	0.12*	0.19*
Q	Fellow professionals outside ISRO	0.94*	79.64*	47.04*	35.27*	0.08	0.18*
R	Results of own experiments	0.95*	77.19*	51.78*	40.37*	0.07	0.24*
S	Professional Meetings seminars etc.	–0.98*	70.00*	77.10*	30.94	0.12*	0.36*
T	Educational and training courses	–0.18*	43.81*	33.69	11.73	0.11*	–0.09

Key : * The association/correlation is statistically significant at 0.05 significance level *(p)* for a two tailed test.

Note : The tabulated values of x^2 at 0.05 significance level *(p)* for a two tailed test for 16, 24 and 20 degrees of freedom *(df)* are 26.296, 36.415 and 31.410 respectively.

TABLE 7.9 : Dependence on Informal and Interpersonal Sources of Information Vs Status

(Weighted Mean Score)

Informal and Interpersonal Sources of Information	Status										
	A,B,C & D	E	F	G	H	I	J	K	L	M&N	Total	r_s
M Personal experience	3.05	3.41	3.42	3.25	3.04	3.02	2.71	2.95	2.81	2.23	3.10	0.8788*
N Superiors	1.71	1.94	2.02	2.32	2.38	2.42	2.55	2.90	2.75	2.60	2.33	-0.9515*
O Peers and colleagues in ISRO	2.14	2.32	2.16	2.11	1.98	1.94	1.94	1.95	2.23	1.87	2.08	0.5727
P Subordinates and juniors	2.09	2.22	1.90	1.69	1.41	1.49	1.48	1.49	1.29	1.62	1.72	0.7121*
Q Fellow professionals outside ISRO	1.81	1.74	1.59	1.37	1.43	0.93	0.88	1.00	0.53	0.83	1.27	0.9394*
R Results of own experiments	2.90	3.06	3.05	2.87	2.73	2.30	1.97	1.85	1.91	1.50	2.57	0.9515*
S Professional meetings, seminars etc.	2.24	2.20	2.20	1.89	1.79	1.48	1.45	1.05	1.13	0.88	1.75	0.9848*
T Educational and training courses	0.90	1.43	1.68	1.59	2.00	1.68	2.07	2.10	2.10	1.89	1.70	-0.8121*

Key: r_s Spearman rank order correlation; *, The critical value of t at 0.05 significannce level in a two tailed t-test for 9 degrees of freedom is 2.262 and hence correlations marked * are found valid at 0.05 significance level.

Note: Status A to N are in the descending order of hierarchical rank.

p.36). has found. However, dependence on results of own experiments, professional meetings and seminars, and educational and training courses showed almost linear positive relation with qualifications of the users. The doctorates have depended to the maximum on these sources.

The dependence of the space technologists on superiors, fellow professionals outside ISRO, results of own experiments and professional meetings and seminars are also found to be significantly related to the nature of work of the space technologists (Table 7.11). The dependence on the last three informal sources of information has almost linearly increased with increase in management/supervisory content of the job or decrease in operational activity and *vice versa.*

It can be inferred from Table 7.12 that the degree of dependence on informal sources like superiors, subordinates/juniors, fellow professionals outside the organisation and results of own experiments varied significantly with the specialisation of the IST. Electronics and mechanical engineers[10] depended most, and the aeronautical and structural engineers the least on

TABLE 7.10 : Dependence on Informal and Interpersonal Sources of Information Vs Qualifications

(Weighted Mean Score)

	Informal and Interpersonal Sources of Information	Qualifications					Total	x^2
		U	D	B	M	P		
M	Personal experience	2.63	2.87	3.23	3.24	3.54	3.08	43.5621*
N	Superiors	2.59	2.52	2.24	2.22	1.78	2.33	38.8642*
O	Peers and colleagues in ISRO	2.01	1.93	2.12	2.25	2.13	2.10	19.7628
P	Subordinates and juniors	1.65	1.53	1.72	1.89	1.96	1.75	26.0046
Q	Fellow professionals outside ISRO	0.47	0.89	1.29	1.39	2.08	1.17	79.6381*
R	Results of own experiments	1.66	2.28	2.82	2.73	3.18	2.53	77.1922*
S	Professional meetings, seminars, etc.	0.91	1.54	1.17	2.03	2.13	1.67	70.0040*
T	Educational and training courses	1.78	2.14	1.63	1.57	1.26	1.71	43.8051*

Key : *, The association is statistically significant at $p = 0.05$ as the tabulated value of x^2 at $p = 0.05$ for 16 degrees of freedom is 26.296, *Qualifications*: U-Undergraduate, D-Diploma holder, B-Graduate, M-Postgraduate. P-Doctorate.

superiors for information. But dependence on subordinates/juniors is maximum by aeronautical and structural engineers and minimum by physicists. Again, it is aeronautical and structural engineers along with electrical engineers expressed highest dependence and mechanical engineers the least dependence on results of own experiments. Further as found by Shuchman (1981, p. 36) aeronautical and structural engineers together with physicists depended to the maximum on fellow professionals outside the organisation.

The experience of the space technologists has showed (Table 7.8) a slight positive correlation ($r = 0.12$) with dependence on subordinates/ junior and professional meetings and seminars and a slight negative correlation with dependence on superiors ($r = -0.17$) and educational and training courses ($r = -0.11$) as informal sources of information. It is strange that the length of experience of the respondents does not correlate with their dependence on 'personal experience' as an intrapersonal source

TABLE 7.11: Dependence on Informal and Interpersonal Sources of Information Vs Nature of Work

(Weighted Mean Score)

Informal and Interpersonal Sources of Information	Nature of work								
	A	B	C	P	D	E	F	Total	x^2
M Personal experience	3.22	3.20	3.18	3.27	2.82	3.14	2.68	3.07	27.5798
N Superiors	1.67	2.33	2.24	2.61	2.72	2.29	2.15	2.35	68.9712*
O Peers and colleagues in ISRO	1.87	2.29	2.09	2.00	2.21	2.08	1.95	2.12	33.7384
P Subordinates and Juniors	2.02	1.81	1.66	1.54	1.70	1.80	1.52	1.72	24.8980
Q Fellow professionals outside ISRO	1.87	1.39	1.17	1.28	0.95	0.93	0.77	1.18	47.0425*
R Results of own experiments	2.92	2.52	2.71	3.16	2.14	2.35	1.94	2.52	51.7821*
S Professional meetings, seminars, etc.	2.15	2.11	1.74	1.50	1.38	1.03	1.35	1.64	77.1039*
T Educational and training courses	1.71	1.61	1.68	1.37	1.85	2.18	1.27	1.72	33.6851

Key : *, The association is statistically significant at $p = 0.05$ as the tabulated value of x^2 at $p = 0.05$ for 24 degrees of freedom is 36.415; *Nature of Work:* A-Management/ Supervision, B-Planning/Systtem Analysis, C-Design and Development, D-Fabrication and Testing, E-Operational Activity, F-Others.

TABLE 7.12 : Dependence on Informal and Interpersonal Sources of Information Vs Specialisation

(Weighted Mean Score)

Informal and Interpersonal Sources of Information	Specialisation A	C	E	F	H	I	Total	x^2
M Personal experience	3.18	3.55	2.84	3.10	3.00	3.11	3.06	20.3255
N Superiors	2.03	2.06	2.34	1.63	2.29	2.42	2.33	35.6214*
O Peers and colleagues in ISRO	1.97	2.40	1.91	2.05	1.75	2.26	2.11	25.1013
P Subordinates and juniors	1.62	1.89	1.66	1.94	1.88	1.70	1.71	40.6986*
Q Fellow professionals outside ISRO	1.59	1.17	1.00	1.68	0.88	1.16	1.17	35.2715*
R Results of own experiments	2.56	2.74	2.12	2.79	2.79	2.65	2.51	40.3672*
S Professional meetings, seminars, etc.	2.10	1.95	1.39	1.89	1.73	1.68	1.66	30.9437
T Educational and training courses	1.68	1.65	1.65	1.89	1.93	1.70	1.70	11.7348

Key : *, Association is statistically significant as tabulated value of x^2 at $df = 20, p = 0.05$ is 31.410; *Specialisation:* A-Physics, C-Mathematics E-Mechanical Engineering, F-Aeronautical and Structural Engineering, H-Electrical Engineering, I-Electronics.

of information. In other words, mere number of years of experience does not enrich personal experience but increasingly does so with increase in status and qualifications.

Lastly, professional activities and achievements of the IST showed (Table 7.8) low but definite positive correlation with dependence on results of own experiments (r = 0.24) and professional meetings and seminars (r=0.36), a slight positive correlation with dependence on peers and colleagues (r = 0.16), subordinates/juniors (r = 0.19) and fellow professionals outside the organisation (r = 0.18) and a slight negative correlation with dependence on superiors (r = –0.13) for information. It is but natural that as one is more experienced and professionally active, he depends less on superiors for work-related information.

The findings that higher the status and longer the experience, the more is the dependence of the IST on subordinates and juniors for information could be as part of the delegation of information-gathering work to juniors in order to involve them in team work checked by 73.7 per cent of those delegated (Chapter 6).

7.5 Relative Dependence on Formal and Informal Sources of Information

It is clear from Table 7.13 that a little less than half (nearly 45 per cent) of the respondents felt that they depend equally on both formal and informal sources. In addition, about one-fourth (nearly 24 per cent) depend more on formal sources than informal sources and 4.4 per cent depend mostly on formal sources. Hence, substantial dependence on formal and informal sources of information are, respectively, by 28.2 per cent (23.8+4.4) and 26.4 per cent (17.2+9.2). In other words, a little less than half of the respondents lay equal emphasis on both formal and informal sources (C), little over one-fourth (26.4 per cent) indicate a slant towards informal sources and the rest (28.2 per cent) show a slant towards formal sources. The data do not fully support the general findings of DoD users study (Auerbach, 1966, p. 106-107; 1965, p. 1-12), Raitt (1984, p. 209, 211 and 214) and Shuchman (1981, p. 30,53-55) that users depended heavily on informal sources. There are 86 (17.2 per cent of) respondents in the present study who depend almost on informal sources only and another 46 (9.2 per cent) who depend more on informal sources of information than formal sources. The data, therefore, indicate that there are more space technologists who exclusively depended on informal sources of information than formal but the response population as a whole has slightly slanted towards formal sources of information.

TABLE 7.13: Relative Dependence on Formal and Informal Sources of Information

	Relative Dependence	No.	%
A.	Almost on informal sources only	86	17.2
B.	More on informal sources than formal sources.	46	9.2
C.	Equally on both informal and formal sources.	226	45.3
D.	More on formal sources than informal sources.	119	23.8
E.	Almost on formal sources only.	22	4.4
	Total	499	99.9

Note: 1. Eighteen not answered and eighteen invalid responses are excluded.

2. The number of anonymous respondents for options A to E respectively are 5, 2, 15, 6, 1.

7.6 Relation of Relative Dependence on Formal and Informal Sources of Information with User-characteristics

Table 7.14 presents the results of chi-square test to see the association of relative dependence of the space technologists on informal sources and formal sources with the six user-characteristics.[11] From the table it is clear that all the characteristics except experience of the users show statistically significant relation with their relative dependence on formal and informal sources of information. The individual contingency tables on which chi-square statistic calculations have been made are presented as Tables 7.15, 7.16, 7.17, 7.18, 7.19 and 7.20.

An examination of Table 7.15 reveals that, by and large, higher the status, the more is the dependence on formal sources and vice versa. Similarly, Table 7.16 shows that higher the qualifications, more is the dependence on formal sources of information and vice versa. Further, the management/supervision, planning/system analysis and design and development personnel have depended more on formal sources and fabrication, and testing personnel depended more on informal sources of information (Table 7.17).

Physicists and aeronautical and structural engineers have depended more on formal sources of information and mechanical engineers on informal sources of information (Table 7.18) than others. The experience of a seeker of informaion showed no significant relation with relative

TABLE 7.14: Results of Association Test of Relative Dependence on Informal and Interpersonal Sources compared to Formal and Documentary Sources of Information with User-characteristics

User-characteristics	Degrees of Freedom	Tabulated Value of x^2 at 0.05 Significance level	x^2
Status	36	43.773	134.78*
Qualifications	16	26.296	125.48*
Nature of Work	24	36.415	72.14*
Specialisation	20	31.410	38.13*
Experience	16	26.296	17.74
Professional Activities and Achievements Index	8	15.507	25.68*

Key: *, The asociation is statistically significant at $p = 0.05$

dependence on formal and informal sources of information (Table 7.19). However, those who depended almost exclusively on formal as well as informal sources of information are slightly more experienced than others. But from the personal observations of the investigator and what the respondents have expressed in answer to the questionnaire, it appears that new entrants have depended more on library/literature in general. However, for those working on the projects, internal reports and senior colleagues have been two additional important sources. To quote one, "since I am a new entrant to this organisation I am highly depending on library for information. I also depend on my colleagues and seniors for information relating to projects undertaken at ISAC". Another says, "I mostly depend on various documents brought out on IRS. At the same time I am brushing up my mind with some text books on optics and electronics".

There is a very clear, significant, positive and linear relationship of professional activities and achievements of the space technologists with their relatively higher dependence on formal sources of information (Table 7.20). For instance, those who depend almost on informal sources have a mean professional activities and achievements index value of 4.34 as against 16.25 of those depending almost on formal sources. The dependence on formal sources of information and professional activities

TABLE 7.15 : Relative Dependence on Informal and Interpersonal Sources Compared to Formal and Documentary Sources of Information Vs Status

	Relative Dependence					Total
Status Code	**A**	**B**	**C**	**D**	**E**	
A B C & D	0	2	7	9	2	20
E	2	4	19	22	4	51
F	1	8	29	22	5	65
G	8	12	52	28	4	104
H	7	8	22	11	3	51
I	14	5	28	11	1	59
J	15	4	27	12	0	58
K	12	0	8	2	0	22
L	14	1	21	2	1	39
M & N	12	1	5	0	0	18
Total	85	45	218	119	20	487

($x^2 = 134.778$, $df = 36$, $p < 0.05$)

Note : Status codes A to N are in the descending order of hierarchical rank.

Key : A-almost on informal sources B- more on informal sources than formal sources, C-equally on both informal and formal sources, D-more on formal sources than informal sources, E-almost on formal sources.

TABLE 7.16 : Relative Dependence on Informal and Interpersonal Sources Compared to Formal and Documentary Sources of Information Vs Qualifications

		Relative Dependence					Total
Qualifications		**A**	**B**	**C**	**D**	**E**	
U	Undergraduate	42	3	34	4	1	84
D	Diploma holder	19	6	48	16	1	90
B	Graduate	14	17	65	40	10	146
M	Postgraduate	8	16	67	46	5	142
P	Doctorate	0	2	5	12	4	23
	Total	83	44	219	118	21	485

($x^2 = 125.4811$, $df = 16$, $p < 0.05$)

Key: A-almost on informal sources, B-more on informal sources than formal sources, C-equally on both informal and formal sources, D-more on formal sources than informal sources, E-almost on formal sources.

TABLE 7.17 : Relative Dependence on Informal and Interpersonal Sources of Information Compared to Formal and Documentary Sources of Information Vs Nature of Work

		Relative Dependence					Total
Nature of work		**A**	**B**	**C**	**D**	**E**	
A	Management/Supervision	3	4	14	16	3	40
B	Planning/System Analysis	6	6	21	19	4	56
C	Design and Development	17	19	62	50	8	156
D	Fabrication and Testing	43	7	63	13	2	128
E	Operational Activity	3	1	19	4	2	29
F	Others	6	0	9	4	1	20
P	Both C and D	3	5	13	6	0	27
	Total	81	42	201	112	20	456

($x^2 = 72.1371$, $df = 24$, $p < 0.05$)

Key : A-almost on informal sources, B-more on informal sources than formal sources, C-equally on both informal and formal sources, D-more on formal sources than informal sources, E-almost on formal sources.

TABLE 7.18 : Relative Dependence on Informal and Interpersonal Sources of Information Compared to Formal and Documentary Sources of Information Vs Specialisation

Specialisation		Relative Dependence					
		A	B	C	D	E	Total
A	Physics	1	2	16	19	2	40
C	Mathematics	1	3	11	4	0	19
E	Mechanical Engg.	32	10	51	24	5	122
F	Aeronautical and Structural Engg.	1	2	5	9	2	19
H	Electrical Engg.	2	3	8	2	1	16
I	Electronics	38	21	98	47	9	213
	Total	75	41	189	105	19	429

($x^2 = 38.1326$, $df = 20$, $p < 0.05$)

Key : A-almost on informal sources, B-more on informal sources than formal sources, C-equally on both informal and formal sources, D-more on formal sources than informal sources, E-almost on formal sources.

TABLE 7.19 : Relative Dependence on Informal and Interpersonal Sources of Information Compared to Formal and Documentary Sources of Information Vs Experience

Relative Dependence	Experience in Years						
	<3	3-5	6-10	11-15	>15	Total	Mean
A	7	9	21	10	7	54	9.30
B	4	9	15	6	1	35	7.46
C	32	27	57	44	13	173	8.48
D	13	16	31	22	12	94	9.45
E	3	2	3	9	1	18	9.72
Total	59	63	127	91	34	374	8.80

($x^2 = 17.7399$, $df = 16$, $p > 0.05$)

Key : A-almost on informal sources, B-more on informal than formal sources, C-equally on both informal and formal sources, D-more on formal than informal sources, E-almost on formal sources.

TABLE 7.20: Relative Dependence on Informal and Interpersonal Sources of Information Compared to Formal and Documentary Sources of Information Vs Professional Activities and Achievements Index

Relative Dependence	Professional Activities and Achievements Index values 0-5	6-15	>15	Total	Mean
A	20	1	1	22	4.34
B	8	3	3	14	11.18
C	41	32	17	90	11.48
D	16	16	15	47	15.60
E	2	5	3	10	16.25
Total	87	57	39	183	11.92

($x^2 = 25.682$, $df = 8$, $p < 0.05$)

Key: A-almost on informal sources, B-more on informal than formal sources, C-equally on book on formal and formal sources, D-more on formal than informal sources, E-almost on formal sources.

and achievements as defined in this study are very strongly related.

7.7 Blending of Formal and Informal Sources of Information: Observations Based on Interviews

There is an interdependence and neat interplay between formal and informal sources of information. During interviews with the space technologists, it was also discussed how far the formal sources of information supplement the informal sources of information in providing work-related information. The general consensus of the respondents is that exclusive dependence on formal sources is possible and may be adequate in some work situations but exclusive dependence on informal sources might help to pull on with job but certainly result in lack of efficiency, effectiveness and innovative spirit.[12]

Further, the informal channels are generally not information rich *i.e.*, they contain relatively more noise than formal sources.[13] Proportional and appropriate blending of these two channels is very much felt by the interviewees. By and large, such a blending depends on the capacity of the individual concerned and the organisational environment.

Some measures suggested (during the discussion) to make the formal sources supplement the informal and *vice versa* are:

(i) Organisation of internally generated information in the form of reports, drawings, papers, photographs, etc.;
(ii) Inhouse technical journal;
(iii) Expertise data bank to enable getting information quickly on allied areas;
(iv) Component co-ordination group;
(v) Regular inhouse seminars;
(vi) More opportunity for meeting counterparts outside the organisation;
(vii) Inviting people from both inside and outside the organisations for presentations in the organisation (which not only makes both speaker and audience to keep informed but also promotes use of formal sources intermittently with informal discussions);
(viii) A positive management policy to create such conducive environment; and
(ix) Journal clubs.

The space technologists have favoured informal sources when information particularly on allied topic has to be obtained quickly and when quick decisions have to be made. On the other hand, formal sources are sought in the circumstances where risk is to be minimised, decisions have to be passed on to higher ups, authenticity of information from informal sources has to be checked and gaps in information available from informal sources have to be filled. Many examples have been cited by the respondents about how they come to know about parallel thinking within the organisation during seminars and discussions which in turn lead them to formal sources. They felt that information from experts is at times difficult to put into action and on the other hand the difficult-to-digest information from formal sources can be got clarified through informal discussions.

A sort of journal club did exist in one of the project core teams where each one of the members used to speak once in a fortnight on his area of work to others in the group. A young engineer strongly felt that regular meetings of group of persons working on the same areas like 'thin films', 'microprocessor based system' and 'remote sensing' but placed in different sections/divisions/projects are necessary.

7.8 Keeping up-to-Date With Latest Developments

Keeping abreast of latest developments in the field is a predominant purpose of seeking information and the IST have highly sought S&T news and basic S & T information. How do they deploy various sources of information for the purpose of keeping up-to-date was discussed during interviews. It was found that over 52 per cent of the respondents used journals, 13 per cent depended on discussion with colleagues and 9 per cent on trade literature to keep themselves up-to-date. Other sources mentioned are seminars, newspapers, preprints, lectures and sales representatives.

The weekly summary of information-related activities also clearly showed that journals form the largest single source for keeping up-to-date about latest developments as long as area is specialised and specific. Shuchman (1981, p. 44). also found that most engineers (70 per cent) read journals to keep abreast in their own field.

However, in the area of Astronomy and Astrophysics, it is preprints and IAU telegrams and in case of those concerned with equipment, test and evaluation, it is trade literature and sales representatives which are important sources of information to keep up-to-date. For generalists, in addition 'o journals, occasional seminars, discussion with colleages, professional lectures and even newspapers help to keep up-to-date.

7.9 Stimuli/Impetus for New Ideas or Novel Thoughts

It was also discussed with respondents as to how they are stimulated to new ideas or novel thoughts. The respondents felt that both discussion with similar minds and conscious reading lead to new ideas in case of over two-third of the respondents. The result is close to what is already known from the past studies. Other stimuli are noncommitted thinking, listening to lectures and even writing.

7.10 Tolerable Delay in Supplying Information and 'Late Detection' of Information

The discussion with the sample space technologists revealed that the delay in supplying specific information, particularly for project personnel, is very costly. On the other hand, when a particular document is needed most of the times substitutes are available and hence the time factor becomes more flexible. By and Large, the tolerable delay for supplying information (and documents) depended on specific instances. The tolerable delay expressed by the respondents ranged from instantaneously to a maximum of one week for specific information and a couple of days to three months for documents.

Many felt that late detection of useful information happens time and again and often it leads to frustration. Such instances of late detection or receipt of crucial information occurred normally when they wrote to manufacturers of equipment, components and materials, when pointed out by colleagues, when detected accidentally by oneself and due to late procurement of documents by the library. The following specific instances are interesting:

(i) Due to late procurement of ASTM references radiographs by the library a respondent could not use the same for comparing radiographs of a spacecraft project with that of ASTM standard to find out acceptance of magnesium casting;
(ii) It was in the halfway through that a project engineer discovered useful information that titanium alloy with gold coating would have been better than aluminium for a particular structural element of spacecraft and he repented that the same could not be implemented in the project on hand due to shortage of time to test the new device;
(iii) Another engineer felt that Harological Department of HMT had a jig boring machine which could have been used from the beginning for laying solar cells at calculated and exact distances before soldering instead of laying the cells by hand. He felt that neither systematic exploration for such information was made nor it was widely disseminated among the concerned.

As compared to earlier findings of user research, the IST have not just ignored late information but are helpless due to time-bound nature of work, yet they are conscious of such instances.

7.11 Technological Gap and Age of Information

It has been discussed with the respondents during interview about the technological gap they are having in their present work compared to the technology of developed countries. Depending on specific field, the technological gap is felt from 5 years to 15 years. A majority of the space scientists felt it as 10 years. Some respondents hasten to add that it does not mean that they do not need or they cannot digest information about current state-of-the-art. They felt the limitation is in gadgets available for dealing with state-of-the-art technology. Some were of the opinion that the technological gap is almost nil in some specialised fields.

Majority felt that the technological gap is being narrowed down over years in their areas and the rest is equally divided between 'no change' and 'widened'. This aspect of bridging the technological gap varied very much from subject to subject. Even where gap is widened or it is alreay too much, it was felt that there is no need for them to go through the same steps as did in the developed countries; they could skip certain stages. Secondly, in some cases the technological gap appears to be widened because of the deviations in the areas of work and hence, it apparently looks that technological gap is widened.

As part of the above discussion the respondents were asked about the age of the documents and information acceptable to them. The rate of obsolescence of information and literature in different areas within space science and technology has varied very widely. Except astronomy, in all other areas document literature older than 20 years are found to be of negligible use. In this connection, it is also pointed out by the respondents that certain aspects of old technology are valid for a very long time (*e.g.* satellite transponder, travelling wave tube amplifier, etc.,) and classified literature of advanced countries do take a longer time to become absolete. However, the technological progress need not always be incremental.

7.12 Satisfaction About Existing Sources of Information

Finally, a discussion about the respondents' satisfaction about existing sources revealed mixed reactions. By and large, they are fairly satisfied with the sources of information presently available to them but there existed a lot of scope to improve upon them. Though it appears as a too general finding, many felt that they should devote more time and efforts for gathering information in a systematic way. They wished they should have put more efforts and spent more time on systematic information gathering than what they are doing at present. This reinforces the finding of Raitt (1984, p. 228). To quote an engineer 'I feel I could have devoted more time for information collecting activities to do better". Another senior engineer states that he is not satisfied with the existing sources of information and more seminars and journal clubs have to be organised.

NOTES

1. The dependence of the space technologists on journals showed low but definite and positive correlation ($r = 0.25$) with the need for S & T news and theoretical

background/basic S and T information.

2. As there is a low but definite positive correlation (r =0.25) between the number of internal reports prepared during the last one year and the degree of dependence on official documents and internal reports, those who prepare internal reports tend to depend slightly more on them.
3. Since a low positive correlation (r = 0.34) is found between the degree of dependence on reprints and preprints and the number of papers published, one can conclude that the more a respondent publishes papers, the more he depends on reprints and preprints.
4. The requirement of the standard and patent specifications as a type of information (which ranked 9th among types of information sought by the IST vide Table 5.1) and the degree of dependence on standard and patent specifications (which now ranked as 10th) is moderate and positive (r = 0.57).
5. The 'other informal sources' of information mentioned are 'sales representatives' (2), 'news media' (3) and 'discussions during tours and seminars' (1).
6. 'Results of one's/own experiments' which has ranked as the second informal source of information for the space technologists, has an almost moderate correlation (r = 0.35) with 'checking and evaluation of one's/own results' and 'orienting one's/own work with the existing body of knowledge' (r = 0.34) as the purposes of seeking information.
7. The dependence on professional meetings, seminars, etc. for information correlated moderately and positively with the dependence of respondents on conference proceedings and papers (r = 0.49) and seeking information for the purpose of participating in seminars and conferences (r = 0.48). Further a low and positive correlation (r = 0.31) is found between number of professional meetings, seminars attended during the past one year by the IST and their dependence on them as an informal source of information.
8. The dependence on training and educational courses related significantly to part-time studies undertaken (x^2 = 16.93, df = 4, p < 0.05) but not to inservice trainings received (x^2 = 4.32, df = 4, p > 0.05).
9. A slight positive correlation (r = 0.17) is found between the degree of dependence on fellow professionals outside the organisation and the number of tours performed by the respondent during the past one year.
10. Shuchman (1981, p. 36). also found that 'the supervisor is an important information source for mechanical engineers'.
11. There is no significant difference between male and female space technologists in their relative dependence on formal and informal sources of information (Sridhar, 1987).
12. Even then there are 17.2 per cent respondents depending almost on informal sources as against 4.4 per cent depending almost on formal sources (Table 7.13).
13. The Indian space technologists devote less time to informal sources than formal sources.

8

THE INFORMAL COMMUNICATION NETWORK AND THE COMMUNICATION BEHAVIOUR OF THE INDIAN SPACE TECHNOLOGISTS

8.1 Inter-personal Information Sharing among the Indian Space Technologists

The extent to which the Indian space technologists share information is presented in Table 8.1 based on data gathered from the questionnaire (Q.No. 9). The respondents were asked to mark on a four-point scale the extent to which the other space technologists share information with them. The response was sought in relation to the four groups as shown in the Table.

The number of the respondents who felt that 'no information is shared' is 106 in case of 'others' (*i.e.*, other than A, B and C in the Table 8.1), 45 in case of subordinates and juniors, 28 in case of superiors and a negligible 12 in case of peers and colleagues. The number of respondents who felt that 'not much information is shared' (code 1) by subordinates and juniors, peers and colleagues, superiors and others are respectively, 72, 56, 85 and 94. As per the mean score shown in Table 8.1, peers and colleagues rank first in the degree of sharing information, followed by superiors, subordinates and juniors and 'others'. Thus there exists greater and more free inter-personal communication and information-sharing among peers and colleagues than among subordinates and superiors.[1] Adding/subtracting one standard deviation to/from mean, one can say that nearly 68 per cent of the respondents felt that their peers and colleagues, subordinates/juniors and superiors share information to the extent of 1.45-2.97, 1.02-2.90, and1.14-2.84 (on a four point scale),

TABLE 8.1 : Interpersonal Information-Sharing among Individuals within ISAC

Interpersonal Sources		Not Shared(O)		Extent of Information Sharing					Mean	S.D.
							Total			
		No.	%	1	2	3	No.	%		
A	Subordinates and juniors	45	10.0	72	186	145	403	90.0	1.96	0.94
B	Peers and colleagues	12	2.6	56	214	177	447	97.4	2.21	0.76
C	Superiors	28	6.1	85	213	136	434	93.9	1.99	0.85
D	Others	106	29.4	94	113	47	254	70.6	1.28	1.02

Key : O, No information is shared; 1, Not much information is shared; 2, Certain type of information is shared; 3, Most of the information is shared.

Note : 1. Seventy two, 62, 57 and 170 respondents have not answered A,B,C and D respectively and invalid responses varied from 4 to 16.

2. The mean values of anonymous responses to A,B,C and D respectively are 2.00, 2.12, 1.85 and 1.29.

respectively.[2] The histograms in Fig. 8.1 depict the tendencies discussed above in respect of inter-personal information-sharing.

A free flow discussion with selected respondents during the interview showed some inhibitions and conditions that existed (more or less covertly) for free inter-personal flow of information. The major factors that inhibit sharing of information are lack of mutual trust, confidence and rapport, professional jealousy, potential threat arising out of competition, lack of motivation, team spirit and group discussions, compartmentalised organisational structure, the status consciousness and the fear of getting ridiculed about ignorance on the part of the seeker of information. One engineer proposed how to overcome inhibition to information communication in the following words:

> "A great deal of improvement results if one overcomes the inherent hesitation to admit ignorance. By doing so, you are not only gathering information in a fairly easy manner but also exploiting the human sources of information optimally and without running into the possibility of displeasing them".

A lot of tactics, diplomacy, maneuvering, force and perseverance are needed in extracting information from inter-personal sources. A typical

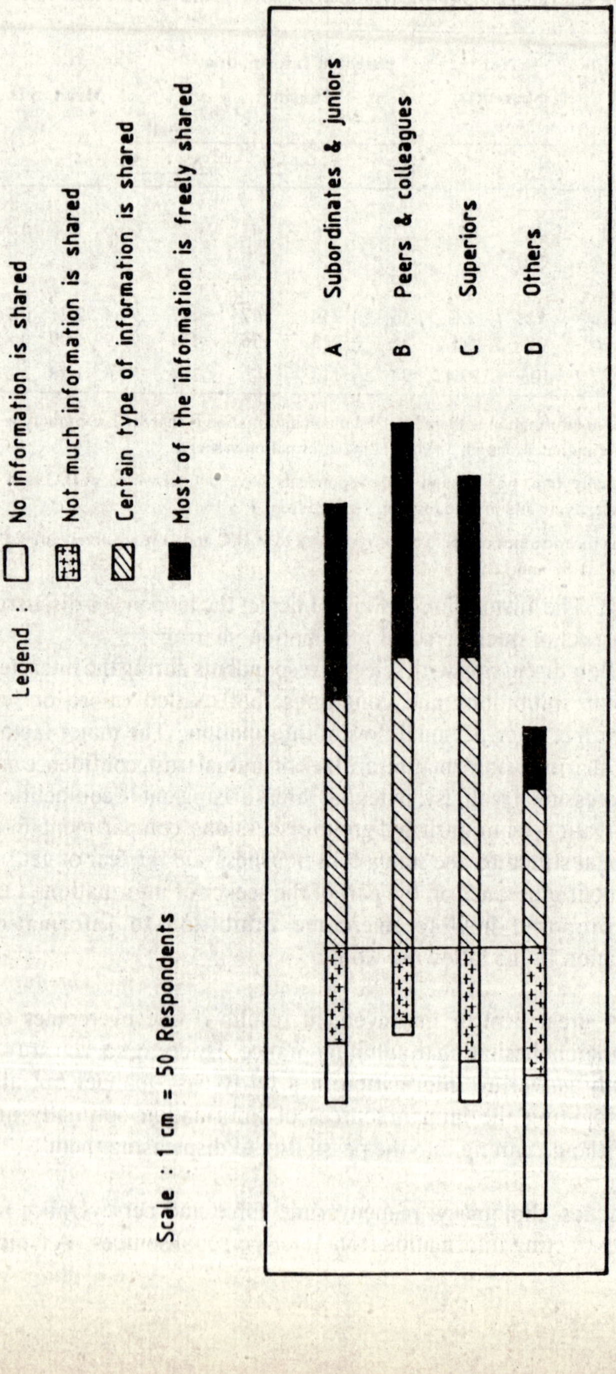

Fig. 8.1 : Interpersonal Information-Sharing among Individuals within ISAC

engineer frankly admits and says "I do withhold information on selective basis". It is also found that information sharing is dependent on content (information), context (time and space) and personality (the person with whom information is shared). For example, an amusing type of information is shared more freely when there is an opportunity to further the interest of the person sharing and when he is sure that the interest of the receiver of information is momentary. Similarly, pure science or basic S&T information and S&T news are more freely shared than applied science and 'how-to-do' information. On the other hand, popular information which has a direct bearing on outlook of people, information which impresses others and gains recognition, competitive information on which two or more parties are working, data pertaining to official routine works are likely to be hidden.

People normally provide information only when somebody asks for it and when they can do it without much extra efforts. To quote a respondent, " when approached, people will give information if they can readily do so. But they are not likely to take extra trouble to give information". The chances of sharing information increases when a seeker of information has a better social relation with a giver (and not being a competitor), when giver wants to show off, gets acknowledgement, believes that the receiver can give something in return,[3] and is interested in the outcome of the application of the information provided. In fact, information-sharing has always been a two-way traffic and everyone wishes to interact with a person of higher information-potential so as to gain more than what one gives away.

It is an interesting issue to probe into the reliance a receiver of information places on information received from inter-personal sources of information. The reliance depends on both the nature of information and the relation between the giver and the seeker of information. It is felt that, generally, the information obtained by inter-personal sources are less reliable but it saves time and provides a trigger for further thinking. Information on a totally new area, and information dealing with inter-disciplinary areas tend to be more relied upon by the receivers. In fact, people with the same specialisation or background tend to share less information.

The actual process of information-sharing with others starts normally with a telephone call and continues in a face-to-face discussion. One's experiments and experiences are related with informal and formal communication in many ways as reflected in the following statement of an engineer:

> "Often, after conducting the tests the results are sent to the persons who are working in the field and their comments/ opinions are sought. They usually send the published literature on that work in addition to their comments."

Yet another says that he collects information generally from literature and discusses it with group members to generate ideas for better application of information. Apart from discussion with colleagues, participation in review meetings of other departments is considered an important way of knowing about other subsystems of satellites.

The degree of dependence for information (Chapter 7) related much better with the degree of information-sharing in case of subordinates and peers than superiors.[4] Those who seek information with the purpose of sharing it with others in the team (Chapter 4) tend to feel that their subordinates freely share information and not the superiors.[5] It appears there exists a lack of confidence among the respondents in free sharing of information by their superiors, eventhough many expressed their dependence on their superiors for information and even the superiors claimed that they collect information for the purpose of sharing it with others in the team.[6]

8.2 Correlation of Inter-personal Information-sharing With User-characteristics

The results of correlation/ association tests of inter-personal information-sharing with the six user-characteristics are shown in Table 8.2. The contingency tables of inter-personal information-sharing against status, qualifications, nature of work and specialisation are respectively presented in Tables 8.3, 8.4, 8.5 and 8.6.

It is clear from Table 8.2 that the professional activities of the respondents are correlated with the degree to which their subordinates, peers, superiors and others share information. In addition, the degree of information-sharing by subordinates varied with status and qualifications of the respondents and information-sharing by peers varied with qualifications of the respondents.

Tables 8.2 to 8.6 reveal that the higher the status (r_s = 0.95), qualifications, management/supervision content of the job, professional activities and achievements (r=0.30) and more the experience of the IST, the more the information-sharing by their subordinates and juniors.

As could be seen from Table 8.4, increased with the qualifications the

TABLE 8.2 : Association/Correlation Test Results of Interpersonal Information-Sharing with User-characteristics

Interpersonal Sources		Status (r_s) $df = 8$	Qualifications (x^2) $df = 12$	Nature of work (x^2) $df = 18$	Specialistation (x^2) $df = 15$	Experience (r) $df = 340$	Professional Activities Achievements Index (r) $df = 173$
A	Subordinates and juniors	0.95*	69.86*	34.29*	17.66	0.19*	0.30*
B	Peers and colleagues	0.58	41.50*	27.79	19.31	-0.06	0.18*
C	Superiors	0.10	20.79	18.70	16.78	0.01	0.17*
D	Others	0.42	17.62	17.59	27.50*	-0.05	0.12*

Key : *, The association/correlation is statistically significant at 0.05 significance level *(p)* for a two tailed test.

Note : The tabulated values of x^2 at 0.05 significance level *(p)* for a two tailed test for 12, 18 and 15 degrees of freedom *(df)* are 21.026, 28.869 and 24.996 respectively.

TABLE 8.3 : Status Vs Interpersonal Information-Sharing (Mean Score)

Interpersonal Sources		Status										
		A,B,C & D	E	F	G	H	I	J	K	L	M & N	Total	r_s
A	Subordinates and juniors	2.67	2.62	2.26	2.13	1.68	1.74	1.60	1.50	1.65	1.08	1.97	0.9515*
B	Peers and colleagues	1.83	2.37	2.29	2.31	2.23	2.25	2.08	1.60	2.10	1.75	2.19	0.5758
C	Superiors	2.05	1.98	1.94	2.13	1.86	2.08	1.86	2.05	2.20	1.67	2.00	0.1030
D	Others	1.13	1.50	1.39	1.29	1.27	1.25	1.23	0.63	1.44	0.91	1.26	0.4182

Key : r_s Spearman rank order correlation; *, correlation is valid at 0.05 significance level

Note : Status A to N are in the descending order of hierarchical rank.

TABLE 8.4: Qualifications Vs Interpersonal Information-Sharing (Mean Score)

Interpersonal sources	Qualifications					Total	x^2
	U	D	B	M	P		
A Subordinates and juniors	1.50	1.75	2.03	2.17	2.63	1.96	69.8563*
B Peers and colleagues	1.82	2.15	2.34	2.27	2.59	2.22	41.5031*
C Superiors	1.86	3.40	1.87	2.16	2.19	1.99	20.7911
D Others	1.06	1.27	1.33	1.24	1.77	1.26	17.6167

Key : *, The association is statistically significant at p = 0.05 as the tabulated value of chi-square at p = 0.05 for 12 degrees of freedom is 21.026; Qualifications-U, Undergraduate; D, Diploma holder; B, Graduate; M, Postgradu ate; P, Doctorate.

Note : The contingency tables on which chi-square test is made are not shown.

respondents have increasingly felt that their peers and colleagues share information. There is also a slight positive correlation of professional activities and achievements of the space technologists with the degree of information-sharing by their peers/colleagues ($r = 0.18$) and superiors ($r = 0.17$). In other words, the more a person is professionally active the more he believes and poses faith in inter-personal information sharing by his subordinates, peers and superiors.

The managers/supervisors and design and development personnel have felt that their subordinates and juniors share information more freely than others.

TABLE 8.5 : Nature of Work Vs Interpersonal Information-sharing (Mean Score)

Interpersonal Sources	Nature of work							Total	x^2
	A	B	C	P	D	E	F		
A Subordinates and juniors	2.41	1.98	2.07	2.13	1.71	1.73	1.50	1.95	34.2887*
B Peers and colleagues	2.36	2.31	2.33	2.18	2.08	2.00	1.94	2.22	27.7881
C Superiors	1.77	1.92	2.13	2.08	1.94	1.69	1.93	1.98	18.7024
D Others	1.45	1.12	1.35	1.00	1.19	1.25	1.72	1.27	17.5854

Key : *, Association is statistically significant at $p = 0.05$ as the tabulated value of chi-square at $p = 0.05$ for 18 *df* is 28.869; *Nature of work*-A, Management Supervision; B, Planning/System Analysis; C, Design and Development; P, Both C and D; D, Fabrication and Testing; E, Operational Activity; F, Others.

Note : The contingency tables on which chi-square test is made are not shown.

TABLE 8.6 : Specialisation Vs Interpersonal Information-sharing (Mean Score)

Interpersonal Sources		Specialisation A	C	E	F	H	I	Total	x^2
A	Subordinates and juniors	2.11	2.25	1.79	2.39	2.33	1.97	1.97	17.6643
B	Peers and colleagues	2.31	2.56	2.06	2.56	2.00	2.28	2.23	19.3126
C	Superiors	2.16	2.50	1.92	2.28	1.93	1.95	2.00	16.7805
D	Others	1.57	1.40	1.03	1.33	1.36	1.37	1.28	27.5006*

Key : *, Association is statistically significant as the tabulated value of chi-square at $p = 0.05$ for $df = 15$ is 24.996; Specialisation: A-Physics; C-Chemistry; E-Mechanical Engineering; F-Aeronautical and Structural Engineering; H-Electrical Engineering; I-Electronics.

Note : The contingency tables on which chi-square test is made are not shown.

8.3 Informal Communication of the Indian Space Technologists

8.3.1 Intra- and Inter-Organisational Communication

The Indian space technologists were also asked through the questionnaire (Q.No. 8.1) to give an approximate number of persons they generally contact regularly for information both within ISRO and outside ISRO. Responses are tabulated in Table 8.7. The frequency distribution shows 10 as the mode number of persons contacted within ISRO and 5 outside ISRO. Naturally, the mode is in the range of 11-15 under the column 'total number of persons contacted' in the Table 8.7. It is also clear from the table 8.7 that a contact of upto 5 persons (both within ISRO and outside ISRO) are made by 27 per cent and upto 15 persons by 65 per cent of the Indian space technologists. As the cumulative percentages in the table 8.7 show a contact of more than 25 persons within ISRO and outside ISRO are respectively maintained by about 8 per cent and 4 per cent of the respondents. A considerable number (*i.e.*, 94 or 27.5 per cent) of the respondents do not contact persons from outside ISRO for information and it corraborates well with the fact that 29.4 per cent of the respondents said that they do not consider fellow professionals outside ISRO as a source of information (Chapter 7, Table 7.7). The average number of contacts per person within ISRO (11.8) is almost double that outside ISRO (6.6).

In the second part of the above-discussed question (*i.e.*, Q.No. 8.2), the respondents were requested to name the five most-often-contacted persons. As many as 391 (out of 535) have provided names of 515 indivduals/organisations and mentioned them 1538 times as most-frequently chosen informal sources of information.[7] The break-up of the number of individuals and organisations contacted with the score of number of respondents mentioned is shown in Table 8.8 and in Fig. 8.2. It is clear that 87 per cent of the contacts are within the organisation (*i.e.*, ISRO) and inter-organisational communication is represented by a meagre 13 per cent. The space technologists being practitioners have maintained a very high intra-organisational communication (Wilkin, 1981, p. 2.10; Pruthi and Nagpaul, 1978, p. 55) and it is believed to be directly related to the performance (Allen and Cohen, 1970, p. 12). The inter-organisational communication is lower than that found by Raitt[8] (19 per cent, 1984, p. 213-214).

A frequency distribution of organisations within the country and outside the country mentioned by the respondents is given in Table 8.9. The contacts of the respondents, though limited within other centres, units and projects of ISRO and moderate with organisations abroad, are widespread over 58 different institutions within the country. As shown in Appendix 6, a very large number of contacts are made with organisations or their wings situated within Bangalore city. Secondly, the number of contacts with 15 academic institutions within the country is alone almost equal to the number of contacts made with the remaining 43 organisations *i.e.*, research organisations, industries, business firms, and other organisations within the country. However, it is noticed that contacts are more with research organisations than industries and with industries than business firms. The maximum contacts of 47 with IISc and 10 with NAL are incidentally the institutions whose libraries are directly used by the Indian space technologists (the analysis of the same is presented later). Thirdly, almost all the business firms contacted are computer or electronics related firms mainly for the purpose of product-information. Fourthly, the educational institutions contacted are mainly the alma maters of the space technologists. Lastly, the large majority of the rest of the organisations within the country, have projects sponsored by ISRO or joint projects with ISRO or got ISRO as their customer or are themselves user agencies/ customers of ISRO. Among the organisations contacted abroad, NASA and ESA (with their subsidiaries) are significant.

8.3.2 *Correlation of Intra - and Inter - Organisational Communication with User-characteristics*

The results of association/correlation tests of a number of regular informal contacts (both within the organisation and outside-the-organisation) with

TABLE 8.7: Frequency Distribution of Number of Persons Contacted as Informal Sources of Information

No. of persons contacted	Within ISRO			Outside ISRO			Total		
	No.	%	Cum. %	No.	%	Cum. %	No.	%	Cum. %
00	4	1.2	1.2	91	27.5	27.5	NA	—	—
01	2	0.5	1.7	10	2.9	30.4	2	0.5	0.5
02	32	7.5	9.2	59	17.3	47.7	30	6.7	7.2
03	32	7.5	16.7	18	5.3	53.0	26	5.8	13.0
04	29	6.8	23.5	23	6.7	59.7	22	4.9	17.9
05	65	15.3	38.8	44	12.9	72.6	42	9.4	27.3
06	24	5.6	44.4	1	0.2	72.8	22	4.9	32.2
07	12	2.8	47.2	1	0.2	73.0	21	5.0	37.2
08	9	2.1	49.3	3	0.9	73..9	11	2.5	39.7
09	1	.2	49.5	1	.2	74.1	12	2.7	42.4
10	92	21.6	71.1	41	12.0	86.1	29	6.5	48.9
11-15	41	9.6	80.7	8	2.3	88.4	73	16.4	65.3
16-20	23	5.4	86.1	16	4.7	93.1	34	7.6	72.9
21-25	24	5.6	91.7	9	2.6	95.7	40	9.0	81.9
26-30	7	1.6	93.3	6	1.8	97.5	17	3.8	85.7
31-35	0	0	93.3	0	0	97.5	14	3.2	88.9
36-40	0	0	93.3	1	0.2	97.7	9	2.0	90.9
41-50	19	4.5	97.8	2	0.5	98.2	12	3.0	93.9
51-60	1	0.2	98.0	1	0.2	98.4	10	2.3	96.2
60-70	0	0	0	0	0	98.4	2	0.5	96.7
71-80	1	0.2	98.2	0	0	98.4	6	1.4	98.1
81-90	0	0	98.2	0	0	98.4	0	0	98.1
>90	6	1.4	99.6	4	0.3	98.7	11	2.5	100.6
Total	421	99.6	99.6	339	98.7	98.7	445	100.6	100.6

Key: N. A., Not applicable.

Note: Not answered and invalid responses to the extent of 111 (69+42) for 'within ISRO' and 196 (179+17) for 'outside ISRO' are excluded. In the total column the partial responses to either 'within ISRO' or 'outside ISRO' are not excluded. The number of 'not answered' and invalid responses in total column is 202.

TABLE 8.8: Pattern of Informal Communication Network (Based on Five Most-often Contacted Persons as Informal Sources of Information)

Most-often contacted Informal sources of Information	No.	No.of times cited	Percentage of total citations
Individuals within ISAC	423	1263	82.1
Section/Division/Project staff within ISAC	3 *	39	2.5
Other ISRO/DOS Centres/ Units/Projects	9	36	2.3
Other organisations within the country	58	162	10.5
Organisations abroad	22	38	2.5
Total	515	1538	99.9

Key : *, Refers to three type of general statements, 'Section Staff' 'Division, staff', and 'Project staff'.

TABLE 8.9: Frequency Distribution of Most-often Contacted Organisations Outside ISAC as Informal Sources

Frequency score of citations	Other Centres, Units Projects of ISRO/ DOS	Organisations with in the country	Organisations outside the country
01	2	38	14
02	0	9	4
03	3	3	3
04	1	0	0
05	0	1	0
06	0	3	0
07	3	0	0
08	0	1	1
09	0	1	0
10	0	1	0
47	0	1	0
	9	58	22

Key : Respondents were asked to name individuals with their organisations. Since very few individuals from outside ISAC have been cited by more than one respondents, all such responses have been coded by the name of the organisation of the persons cited.

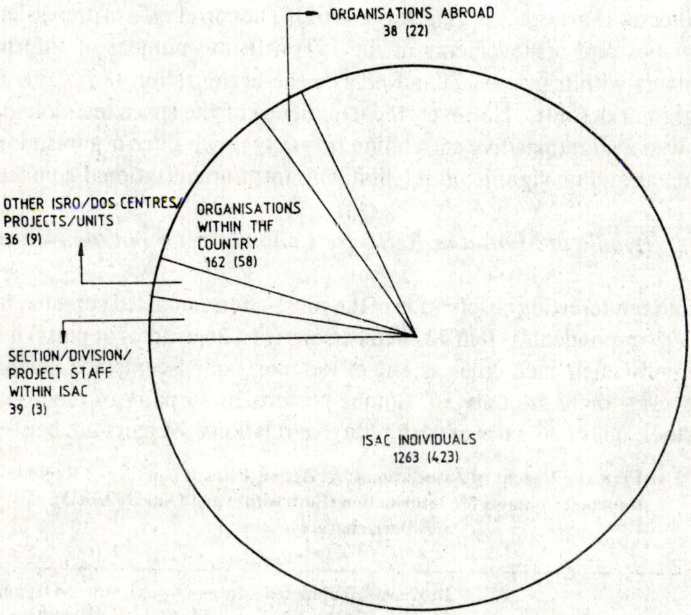

Fig. 8.2 : Composition of Informal Communication Network
(Number of individuals/organisations in bracket)

user-characteristics are shown in Table 8.10. The individual contingency tables of a number of informal contacts versus status, qualifications, nature of work and specialisation of the space technologists are given respectively in Tables 8.11, 8.12, 8.13 and 8.14. It is clear from the tables that there is a linear positive relation of inter-personal informal contacts for information, both within the organisation and outside-the-organisation, to the status, qualifications, management/supervision content of the job and professional activities and achievements of the space technologists.[8] Supervisors had a higher outside-the-organisation contacts in other studies too (Farms, 1972; Hagstorm, 1965). The correlation of professional activities and achievements of the IST with the number of informal contacts within (r = 0.17) and outside-the-organisation (r = 0.20) are slight but definite. However, the experience of the space technologists showed a slight positive correlation (r = 0.14) with inter-organisational contacts and no significant relation with intra-organisational contacts.

8.3.3. *Dyadic or Mutual or Reflexive Communication in the Network*

Another interesting observation of the most-often contacted persons cited by 391 respondents is that 72 cited persons (*i.e.*, 36 dyadics or pairs) have mutually cited each other as one of the most often-contacted persons. However, there are only 56 unique persons in 36 pairs of relations.[10] Further, out of 36 pairs of such reflexive relations, 34 pairs are between

TABLE 8.10 : Results of Association/Correlation Tests of Number of Regular Informal Contacts for Information (Both within and Outside ISRO), with User-characteristics

User-characteristics		Degrees of Freedom	Tabulated value of x^2 at p = 0.05	Intra-Organisational communication	Inter-Organisational communication
Status	(x^2)	14	23.69	71.80*	39.41*
Qualifications	(x^2)	8	15.51	61.27*	42.51*
Nature of work	(x^2)	12	21.03	28.93*	31.68*
Specialisation	(x^2)	10	18.31	16.66	17.93
Experience	(r)	308/261	—	0.07	0.14*
Professional Activities and Achievements Index	(r)	153/132	—	0.17*	0.20*

Key : *, The association/correlation is statistically significant at 0.05 significance level.

persons from the same divisions and of which 32 pairs are between persons of the same sections.[11] Interestingly, ten pairs of persons are of exactly equal status in terms of their designations and of which eight pairs are in the same sections. Out of the remaining 26 pairs of reflexively related persons, the status differences between the persons are, 3 levels in 2 pairs, 2 levels in 11 pairs and just 1 level (or grade) in the rest of 13 pairs.

By assuming that the above 'reflexive relation' satisfies the other conditions of 'equivalence relation', one can decompose the above set of 56 unique respondents into mutually exclusive groups of persons called 'equivalence classes'. Such an analysis has resulted in identifying one class of 9 respondents, another class of four persons and other five classes of two each. This has also given an indication that respondents with codes C0068, C0095, C0110, C0136, C0209, C0255, C0344, C0482, C0565, C0587, C0719, C0768 and C0774 act as linking pins[12] in the first seven classes and the nine respondents of the largest class form a cluster of the network in a limited sense.

8.3.4 *Information-potential of the Space Technologists for Interpersonal Informal Communication*

Now, by looking at the same data from the angle of most-often-contacted

TABLE 8.11 : Number of Regular Informal Contacts for Information Vs Status

Status	No. of persons contacted							
	Within ISRO				Outside ISRO			
	<10	10-19	>19	Total	0	1-9	>9	Total
A,B,C, & D	3	7	8	18	3	5	9	17
E	10	19	6	35	5	14	11	30
F	15	24	16	55	9	22	20	51
G	36	33	19	88	13	44	12	69
H	21	13	8	42	6	22	8	36
I	29	16	8	53	18	16	11	45
J	38	18	2	58	16	22	6	44
K,L,M, & N	55	2	8	65	21	11	8	40
Total	207	132	75	414	91	156	85	332

(x^2 – 71.8024, $df = 14$, $p < 0.05$) ($x^2 = 39.4117$, $df = 14$, $p < 0.05$)

Note ; The status A to N are in the descending order of hierarchical rank.

TABLE 8.12: Number of Regular Informal Contacts for Information Vs Qualification

	Qualifications	No.of persons contacted							
		Within ISRO				Outside ISRO			
		<10	10-19	>19	Total	0	1-9	>9	Total
U	Undergraduate	62	3	6	71	27	9	6	42
D	Diploma holder	46	25	11	82	18	35	17	70
B	Graduate	45	47	32	124	22	56	26	104
M	Postgraduate	47	49	23	119	21	50	29	100
P	Doctorate	5	7	5	17	0	8	8	16
	Total	205	131	77	413	88	158	86	332

($x^2 = 61.27$, $df = 8$, $p < 0.05$) ($x^2 = 42.5126$, $df = 8$, $p < 0.05$)

persons within ISAC (*i.e.* individuals most frequently chosen by the respondents as informal sources of information), the information-potential of persons most often chosen as inter-personal source of information can be derived. Incidentally, as far as inter-personal communication within the population is concerned, seeking as well as communicating information are complimentary to one another. Here, a

TABLE 8.13 : Number of Regular Informal Contacts for Information Vs Nature of Work

	Nature of work	No. of persons contacted							
		Within ISRO				Outside ISRO			
		<10	10-19	>19	Total	0	1-9	>9	Total
A	Management/Supervision	8	14	10	32	5	7	16	28
B	Planning/System Analysis	20	14	11	45	10	18	7	35
C	Design and development	56	51	24	131	23	56	29	108
P	Design, development, fabrication and testing	10	9	3	22	4	9	6	19
D	Fabrication and testing	72	28	14	114	29	36	14	79
E	Operational activity	18	5	3	26	5	16	2	23
F	Others	11	2	3	16	7	5	2	14
	Total	195	123	68	386	83	147	76	306

($x^2 = 28.9297$, $df = 12$, $p < 0.05$) ($x^2 = 31.6849$, $df = 12$, $p < 0.05$).

TABLE 8.14 : Number of Regular Informal Contacts for Information Vs Specialisation

	Specialisation	No. of persons contacted							
		Within ISRO				Outside ISRO			
		<10	10-19	>19	Total	0	1-9	>9	Total
A	Physics	15	14	7	36	3	14	15	32
C	Mathematics	6	6	2	14	4	3	2	9
E	Mechanical engineering	69	27	14	110	26	37	18	81
F	Aeronautical and Structural engineering	4	8	3	15	1	10	5	16
H	Electrical engineering	3	6	3	12	3	4	4	11
I	Electronics	83	58	36	177	41	71	30	142
	Total	180	119	65	364	78	139	74	291

($x^2 = 16.6617$, $df = 10$, $p > 0.05$) ($x^2 = 17.933$, $df = 10$, $p > 0.05$)

respondent mentioning the name of another respondent or any member of the population is treated as a 'relation' in the informal communication network. However, one limitation of the data at this level of analysis is that the official superior-subordinate relations are not excluded.

The frequency distribution of the five most-often- contacted persons cited and the corresponding number of persons citing them is presented in Table 8.15 for individuals within ISAC. Since 423 persons are cited as informal sources by 391 respondents as many as 1263 times (*i.e.,* 1263 dyadic relations), one can see on an average a person is cited 3.23 times by the respondents and the same can be considered to be the average information-potential.[13] In otherwords, on an average, a person cited is an inter-personal source of information for about three (3.23) others in the response population. In a limited sense the connectedness of the informal communication network is 1.57 (*i.e.* 1263 choices or relations divided by 807, the population) and density is 0.6 (*i.e.* 1263 divided by 5 times 423).

8.3.5 *Communication Stars and Technological Gatekeepers*

Assuming that a communication star is one who is contacted as an informal source of information by more than one standard deviation over and above the mean number of contacts by respondents or whose information-potential is atleast one S.D. over and above mean information-potential (*i.e.*, 2.99+3.03 = 6.02 in Table 8.15), there are 44 such communication stars or High-Information-Potentials (HIPs) in the population. These 44 stars forming just 10.2 per cent of the total most-often-contacted persons for information account for 36 per cent of total citations. Compared with general formula that a population of size '*n*' approximately contains $\sqrt{n}$ elites, the star-elites or High-information-potential persons identified is much more than square root of the population (i.e.,$\sqrt{807}$=28.4). Further, excluding 423 who have been cited by one or more of others in Table 8.15, the remaining 384 persons in the population could be considered as 'isolates' or Low-Information-Potentials (LIPs).

Out of 44 stars spotted in this method, 29 are found to have responded to the questionnaire. But only 19 of them have provided upto five names of most-often- contacted persons for information. Out of a total of 88 such citations of names made by these 19 stars, 41 (46.6 per cent) are to other stars, 28 (31.8 per cent) to non-stars within the population and the rest of 19 (21.6 per cent) are to persons outside-the-organisation. As against a 15.3 per cent outside ISAC contacts by the response population as a whole, the communication stars had 21.6 per cent outside contacts. The 'discussion stars' in Hall's study (1972, p.121) have maintained much lower (15 per cent) outside contacts. It is interesting that communication stars have maintained almost half of their contacts among themselves. However, outside the organisation contacts of communication stars are marginally more than others. Assuming that a technological getekeeper is one who links the organisation to external environment six of the 19 respondent-communication stars who had higher than average external contacts for information could be considered technological gatekeepers.

The characteristics of forty-four communication stars as compared to characteristics of the population/response population as a whole are shown in Table 8.16. They differ significantly from others in some respects. The communication stars are characterised by higher age, experience,[14] status, qualifications, professional activities and achievements,[15] use of library and interactions with the library. They are predominantly male managers/planners/technocrats and have acquired promotion more frequently than others. But they do not differ from others in their job satisfaction. Communication stars are twice likely to have some change in their nature of work in recent years than others. The AOCS, 'Facilities' and IGTD divisions appear to have slightly more communication stars than other divisions. Disproportionately more

TABLE 8.15 : Frequency Distribution of Most-often Contacted Persons within ISAC as Informal Sources

Frequency Score of citation	Persons cited			Citations		
	No.	%	Cumulative %	No.	%	Cumulative %
01	180	42.6	42.6	180	14.3	14.3
02	83	19.6	62.2	166	13.1	27.4
03	48	11.3	73.5	144	11.4	38.8
04	33	7.8	81.3	132	10.5	49.3
05	22	5.2	86.5	110	9.0	58.3
06	13	3.1	89.6	78	6.2	64.5
07	5	1.2	90.8	35	3.0	67.5
08	13	3.1	93.9	104	8.2	75.7
09	6	1.4	95.3	54	4.3	80.0
10	4	0.9	96.2	40	3.2	83.2
11	6	1.4	97.6	66	5.2	88.4
12	2	0.5	98.1	24	2.0	90.4
13	1	0.2	98.3	13	1.0	91.4
14	2	0.5	98.8	28	2.2	93.6
15	1	0.2	99.0	15	1.2	94.8
16	1	0.2	99.2	16	1.3	96.1
17	1	0.2	99.4	17	1.3	97.4
18	1	0.2	99.6	18	1.4	98.8
23	1	0.2	99.8	23	1.8	100.6
Total	423	99.8	99.8	1263	100.6	100.6

Notes : 1. Persons from outside ISAC (tabulated in the name of their respective organisations in Table 8.8) are excluded from the present Table.

2. One hundred and thirty-eight not answered, 2 not applicable and 4 invalid responses are excluded.

3. The mean and standard deviation of number of times a space technologist is cited are respectively 2.99 and 3.03.

communication stars are found among physicists and electronics engineers and less among mechanical engineers.

Discussion with selected communication stars and technological gatekeepers revealed that it is part of their duty to collect information from others particularly outsiders and pass on the same to appropriate persons working with them. They also felt that their contacts with information-potential persons (especially those from outside the organisation) are their assets. They appear to read widely on general interest from library books and reports. They have a feeling of self-sufficiency as far as work-related information is concerned.

8.3.6 *Similarities and Dissimilarities of Participants in Inter-personal Communication (Inter-personal Information-potential Vs User-characteristics)*

The relation of some of the characteristics such as technocrat nature, experience, status, specialisations, field of activity, division and section affiliation and qualifications of the space technologists in 1263 dyadic

TABLE 8.16 : Characteristics of Communication Stars

Sl.No.	Characteristics	Population	Communication stars
1.	Average age (in years)	31	36
2.	Males (in %)	92.4	97.7
3.	Average experience (in years)	7.4	11.8
4.	Secured 2 or more promotions (in %)	35.5	75.7
5.	Professional activities and achievements (index)	7.6*	23.8
6.	Status H and above (ie., Gazetted Officers in %)	45.4	95.5
7.	Technocrats (in %)	13.3	100.0
8.	Managers, Supervisors, Planners and System Analysts (in %)	20.5	56.0
9.	Change in nature of work (in % having change)	28.9	55.2
10.	Postgraduates and Doctorates (in %)	31.8	66.6
11.	Library-use (index)	15.4*	21.4
12.	Library-interactions (index)	5.1*	13.7

Key : *, the score is based on in idexes developed and for an average non-star and not for a typical space technologist in the population.

relations of inter-personal contact for information is examined here.[16]

8.3.6.1 ***Technocrat Nature*** : From Table 8.17 one can see that an average technocrat[17] communicates frequently and informally within the population with (6.2) more than double the number of persons an average non-technocrat communicates with (2.6). In other words with inter-personal information-potential of more than six, an average technocrat appears to be a communication star.

8.3.6.2 ***Experience*** : Table 8.18 presents the experience of the space technologists against their inter-personal information-potential in terms of number of persons contacting them for information. The table reveals that 29 persons having 4 to 13 years of experience in the organisation (except one with 18 years of experience) are communication stars (*i.e.*, have information potential of 7 or more) It appears that both the highly experienced as well as the less experienced are not qualified to be communication stars. It is also obvious from the table that the more experienced a person is the more information potential he has ($r = 0.27$, $p < 0.05$). Thus null hypothesis that there is no correlation, other than due to chance, between the number of years of experience (in ISRO) of the space technologist and the number of other space technologists considering him as an informal source for work-related information cannot be completely rejected. On the other hand, some minimum experience appears to be a prerequisite for one to become a 'communication star' but too much of experience appears to be working in the opposite direction. The finding that new recruits do not communicate to many is in agreement with what Gerstberger (1971) has reported, but that the highly experienced do not communicate to many colleagues, is a totally new and unexpected result which needs further study. Table 8.18 also reveals to a limited extent that the number of space technologists to whom another space technologist communicates is curvilinearly related to the experience of the person communicating. In the initial years it appears

TABLE 8.17 : Interpersonal Information-Potential of Technocrats

	No.	No. of times cited	Mean citation
Technocrats	45	279	6.2
Others	378	984	2.6
Total	423	1263	3.0

TABLE 8.18 : Frequency Score (Interpersonal Information-Potential) of Most-often contacted Persons Vs Experience

Frequency Score	Experience in Years																Total
	1	2	3	4	5	6	7	8	9	10	11	12	14	16	17	23	
00	04	01															05
01	17	04		03	02												26
02	06	03	02	01	01												13
03	04	03			02												09
04	09	02			01	01		01									14
05	24	06	07	01	02	01	01										42
06	10	06	04	01	01	01	01	01	01		01						27
07	10	05	02														17
08	02	02		01	01				01								07
09	02	01	04	01		01		02									11
10	20	08	02	03		03	01	02		01	01	01					42
11	04	06		04	01	02		02		01	01				01	01	23
12	01	02	01		02					01		01	01	01			10
13	02		02	02	02		02			01							11
14			01			01											02
15			03	01		01											05
16					01												01
17	01																01
18											01						01
22	01																01
Total	117	49	28	18	16	11	03	10	02	04	04	02	01	01	01	01	268

that he communicates with less number of persons and as the experience exceeds the initial period of say 5 years, he communicates to more colleagues and again declines after say 12 years. In other words, both less experienced and highly experienced technologists communicate to relatively less number of colleagues and those with experience between 6 years and 12 years have communicated to the maximum.

8.3.6.3 ***Status*:** Table 8.19 shows that almost half of the respondents had informal and inter-personal contact for information with persons of higher-status only. A negligible 2.7 per cent and 0.6 per cent had such contacts with lower- status and equal-status persons respectively. On the other hand, the respondents who did not seek informal and inter-personal contacts for information with persons of lower-status, equal-status and higher-status, respectively are 73.9 per cent, 63.3 per cent and 5.5 per cent. The last column of Table 8.19 indicates that nearly three-fourth (72.9 per cent) of dyadic relations of inter-personal contacts are with higher-status persons. The equal-status and lower-status relations are 15.2 per cent and 11.9 per cent respectively. It is not surprising that the IST are status conscious in seeking informal and inter-personal communication. It may be noted that this does not naturally apply to them as communicators. Hence, it can be concluded that the space technologists seem to have considered status as a self-created artificial barrier for free inter-personal communication with colleagues. Inspite of the fact that the IST believed that the peers and colleagues share information more freely than superiors and subordinates, the amount of communication among persons of different status (not necessarily superiors and subordinates) is more than that among persons of the same status, indicating, thereby, high-information-potential of higher status persons.

These findings partly confirm the findings of Allen and Cohen (1970, p. 16) and Arndt et al. (1980), *i.e.*, there is a greater volume of communication from a higher-status persons to lower status persons but not as much as that takes place between equistatus persons. They also support the general finding that information stars and technological gatekeepers are high-status persons and another Indian study (Pruthi and Nagpaul, 1978, p. 55) where more communication flourished among persons of different status than equistatus, but differ from the findings of an European study (Ralli, 1984, p. 254) and other similar abroad studies (Vickery 1973, p. 3 and Havelock et. al. 1969, p. 5.13) that there exists more lateral communication among peers than others.

TABLE 8.19 : Interpersonal Information-potential Vs Status (Relative Status of Most-often Contacted Persons)

Status	Information sought only from		Information not sought from any		Most-often contacted (citing) relations	
	No.	%	No.	%	No.	%
Lower	9	2.7	244	73.9	146	11.9
Equal	2	0.6	209	63.3	186	15.2
Higher	163	49.4	18	5.5	891	72.9
Total	174@	52.7@	471*	142.7*	1223	100.0

Key : *, The categories are not mutually exclusive;

@, Mixed types are not shown and hence total do not tally to 330. However, percentages worked out are based on 330.

8.3.6.4 ***Specialisation and Field of Activity*** **:** As could be seen from Table 8.20, the result appears to be inconclusive as far as relation of field of activity to information-seeking through inter-personal communication is concerned. Of the inter-personal contacts made by the IST, number of persons of different fields of activity is slightly more than number of persons of the same field of activity. On the contrary, more number of persons of the same specialisation are contacted by the IST than persons of different specialisations. Thus the Indian space technologists not only have regular contact for information with colleagues of the same specialisation and field of activity, but also with persons of different specialisations and fields of activity. This, probably, is a reflection of system engineering and multidisciplinary nature of work and a way of cross fertilisation of ideas.

8.3.6.5 ***Divison and Section Affiliation*** **:** The data in Table 8.21 indicate that while little less than half of the informal and inter-personal communications cut-across the official section barrier, only one-fourth of such dyadic communication relations are beyond division barrier. The following inferences can be drawn from the Table 8.21: (i) Intra-divison contacts (74.3 per cent) are almost three times the inter-division contacts (25.7 per cent), (ii) Intra-section contacts (53.6 per cent) are slightly higher than inter-section contacts (46.4 per cent), (iii) Intra-division contacts (74.3 per cent) are considerably higher than the intra-section contacts (53.6 per cent) (iv) Inter-section contacts (46.4 per cent) are nearly double the inter-division contacts (25.7 per cent).

The intra-division/intra-project contacts of 53 per cent found by Raitt (1984, p.163-164) are comparable with the intra-section/intra-project contacts (53.6 per cent) and the inter-section contacts for discussion (28 per cent) found by Bethell (1972, p. 47) is comparable with inter-division/inter-project (25.7 per cent) contacts in the present stduy. The higher number of inter-division contacts was also generally supported by the findings of Shuchman (1981, p. 40).

8.3.6.6. ***Qualifications*** : Since doctorates represented a much smaller fraction (4.5 per cent) of the total population in the present study, the earlier findings of Allen and Cohen (1970, p.16), Bethell (1972, p. 94) and many others that the doctorates choose other doctorates for discussion and nondoctorates tend to contact doctorates, could not be clearly validated. However, out of 56 names cited as most often-contacted persons for information by 13 doctorates, only one-fourth of such contacts of doctorates are with other doctorates and they had a higher rate (30.4 per cent) of contacts outside the organisation than response population as a whole (15.3 per cent) and communication stars (21.6 per cent). The inter-personal information-potential of an average doctorate (5.5) is almost double that found in the population. Out of 32 doctorates, 'communication stars' and 'isolates' are eight each. Doctorate communication stars tend to be technological gatekeepers.

Thus, by and large, the general finding of research in mass-communication that information flows more among homophilies than heterophilies is not true in case of inter-personal communication of work-related information among the IST.

TABLE 8.20 : Interpersonal Information-potential Vs Specialisation and Field of Activity (of Most-often Contacted Persons)

	Specialisation		Field of Activity	
	No.	%	No.	%
Same	325	57	223	45
Different	245	43	272	55
Total	570	100	495	100

TABLE 8.21: Interpersonal Information-potential Vs Division and Section Affiliation (of Most-often Contacted Persons)

	Division of contacting and contacted persons		Section of contacting and contacted persons	
	No.	%	No.	%
Same	913	74.3	653	53.6
Different	316	25.7	566	46.4
Total	1229	100.0	1219	100.0

8.4 Formal Communication Behaviour of the Indian Space Technologists

Before formal communication behaviour of the space technologists is discussed, it may be noted that a large amount of internal formal communication (within the organisation) takes place in what are called semiformal publications *i.e.*, reports and official documents. It is estimated that around 250 such reports are generated every month in ISAC (see Appendix VIII).

About 25 per cent (148) of the sample papers of the space technologists which had 713 citations were examined to find out publishing and citing patterns (Sridhar, 1985). When productivity of papers was examined chronologically over a decade, it was found that there was a weak indication that slightly more papers were published in the year following the year of completion of the major projects of the organisation. The distribution of productivity was highly skewed following the inverse law and 80/20 rule. Almost one-eight of the population or one-fourth of the scientists and engineers in the population had one or more papers to their credit, which is almost the same as that found by Shuchman (25 per cent, 1981) but much lower than that found by Tornudd (35 per cent, 1959). Not only the professional activities and achievements (including publishing activity) highly correlated with the use of the library and user-interactions with the library but also those who published more tend to make more use of the library and their interactions are also more.[18]

Contingency tables of the number of papers published and the number of internal reports prepared against the user-characteristics have shown the following results. Publishing of papers has almost been the activity of persons of higher status (A to G) graduates, postgraduates and doctorates, and persons other than those who are doing operational activity (by nature of work) and has shown a low but definite and positive correlation ($r = 0.34$) with experience of the space technologists. However, preparation of internal reports is a trend found with persons of almost all status (except lowest two status, *i.e.*, M & N), qualifications, nature of work and even those who are less experienced ($r = 0.20$). Thus, as observed already, internal report is a more popular method of communication than formal archival publication like journal articles for the space technologists and it conforms to Shuchman's observation (1982, p.5). It was also found that the percentage of conference papers (33 per cent) brought out was more than that cited (10 per cent) and the percentage of journal articles published (57 per cent) and cited (60 per cent) were almost same. Hence, it may be concluded that the space technologists have a flair for oral conference presentation than archival publication. Further, 46 per cent of the papers were published in Indian journals (only 3 per cent of journal articles cited are of Indian Origin) and nearly 67 per cent of the conference papers were presented at foreign/ international conferences. By and large, both published and cited sources are slanted towards foreign sources, a trend that is more significant in the citations to reports and journal articles. Thus it may be stated that Indian journals served more as publishing outlets than sources for citing. These findings are on line with findings of other Indian studies (Srinivasan, 1970; Shalini and Chudamani, 1978, p.103 and B.N.Singh, 1981, p.179).

The rank order of citations (Sridhar, 1985) to various forms of literature by the IST is: (i) Journals (61 per cent), (ii) Books and monographs (14 per cent), (iii) Reports (10 per cent), (iv) Conference papers and proceedings (10 per cent), (v) Theses and dissertations (2 per cent), (vi) Preprints (2 per cent) and (vii) Lecture notes, standards, product catalogues and private communications (1 per cent). The ranking is very close to what was found in the study of GSFC library (Herner *et al.*, 1979, p. 21) and in another Indian study by Srinivasan (1970) but differs from all other previous studies. One interesting result is that technical reports received 10 per cent citations by the IST as against 3 to 7 per cent in other studies (Burton and Green, Jr., 1961, p.35; B.N. Singh, 1981, p.179; Srinivasan, 1970 and Herner *et al.*, 1979, p. 21).

Defying the general findings that many papers in technology do not carry references to earlier works, only 24 of the 97 papers did not carry references and on an average a paper of the IST carried 7.4 references. But the space technologists have made more self-citations than others, with 35 per cent of the papers having one or more self- citations (*i.e.*, 16 per cent self-citations as against normal 10 per cent self-citations found elsewhere,

Meadows and O'connor, 1969, p.160). Further almost 97 per cent of the citations to journal articles are of foreign origin which is also much higher than what other Indian studies have reported. However, only 79 per cent of the citations to reports are of foreign origin. More than 90 per cent of foreign reports cited are either NASA or ESA reports. Citations to preprints appeared only in the astronomy and astrophysics area. Nearly 91 per cent of journal articles cited are from journals available at ISAC library.

A discussion with respondents about information-production (publishing of papers, reports, etc.), information-transfer and information-consumption revealed the patterns of both formal and informal communication. Publishing activity is found to be more with those associated with TDPs than project or other activities of the Centre. As a corollary, more publications were brought out in allied areas than in the core areas. Compared to information consumption and transfer, information production in terms of papers and other formal publications (except internal reports) is very meagre. It is a matter of common sense that 'more engineers need information than produce it' (Shuchman, 1981, p. 61) and it could also be that the consumption of information by engineers is much more than what they produce. Some engineers have felt that they need encouragement and recognition for writing papers. If quantum of information consumption by an individual is taken as unity, hardly half of it gets transferred to team mates in a digested or condensed way provided the individual happens to be a team leader and takes the responsibility to guide the team. Much less than what one receives, is put into action. Such a role of transferring information is not obligatory and such transfer of information is claimed to be more from a superior to a subordinate than among others.

Respondents who have published papers felt that publishing activity helped them in more than one way in collecting and organising information needed by them. They tend to collect more published-information and cross check authenticity of their own results as well as information collected from different sources. One highly productive author says, "by publishing papers I will be much more cautious compared to oral discussion. I give more importance to authenticity of information collected as well as presented in the paper". Another prolific author narrated how reading and writing of papers trigger each other and help to sustain a temporary zone of interest in him in his search for literature and informal discussion.

An interesting aspect of citation study is how relevant, important and useful are the references cited by the authors. It appears that appending references to a paper is a ritual and it is more so with engineers than scientists. Hence, the extensive use of all the documents cited and their potential use to the reader of paper are in doubt. However, one of the authors said that they do note down references only after using them and such references are highly useful in their work if the theme of the paper is related to their work.

The coauthorship pattern of the IST initially examined for 224 sample papers inclusive of the papers examined for publishing pattern, revealed the following notable findings (Sridhar, 1985): A high coauthorship and hence collaboration was inferred as 81 per cent of the papers had two or more authors *i.e.,* 81 per cent of the papers were multiauthored papers (the ratio of multiauthored papers to total papers, also called 'degree of collaboration' is 0.81); the maximum number of authors per paper was ten; three authored papers were the most frequently occuring papers; the average number of authors per paper (or 'multiplicity index') was 3.17; and the 'productivity ratio'[19] which is inversely related to 'multiplicity index' was 0.32. There existed a very high correlation between collaboration and productivity (Pao, 1980). The IST have about 37 per cent of coauthors from outside-the-organisation.

For the purpose of this study, the sample papers have been further refined by deleting all papers other than journal articles and conference papers, adding some more latest papers (31 journal articles added) and restricting the sample only to papers published by the space technologists while working at ISAC. This sample consisted of 116 papers authored by 104 space technologists (excluding 20 ex-employees of ISAC and 8 outside ISAC authors), had a total of 676 authorships. The three important measures of coauthorship and collaboration, namely, the ratio of multiauthored papers to total papers or 'degree of collaboration' (0.81), the average number of authors per paper or 'multiplicity index' (3.38) and the productivity ratio (0.30) have almost remained the same.[20]

The frequency distribution of the number of coauthors in the sample papers shown in Table 8.22 clearly shows that about 10 per cent of papers have only one author and the remaining 90 per cent have 2 to 15 coauthors. Alternatively, little over one-third of papers have 1 to 5 coauthors and two-third have 6 to 15 coauthors or 10 per cent have 11 to 15 coauthors and 90 per cent have 1 to 10 coauthors.[21]

By assuming that the coauthorship relation satisfies the conditions of equivalence relation (though the transitivity property cannot be strictly proved to be true), we can divide the 104 ISAC authors into nine mutually exclusive or non-intersecting clusters or groups (mathematically, equivalence classes). Table 8.23 presents these groups with their group productivities (of number of papers) based on both full credit as well as fractional credit for multiauthored papers. The table indicates that except groups 1 and 8, no other group has higher productivity than an average of one paper per author. It is in these two groups, nine highly productive authors (with 10 or more papers to their credit) are embraced- eight of them in group 1 and one in group 8. As is generally believed, highly productive authors tended to coauthor with many different persons and became the communication stars in a limited sense. Highly productive authors in group 1 coauthor with 7 to 15 persons. The only highly productive author in group 8 has coauthored with 5 persons.

The coauthorship relations (*i.e.*, pairs or dyads) within ISAC are also examined against the status, section/division/project affiliation, specialisation and field of activity of the space technologists. Nearly 20 per cent (112 out of 562) of coauthorship relations are exactly at the same status and the rest between persons of different status. Further 37, 35 and 18 persons, respectively, have not coauthored with lower status, equal status and higher-status persons. While 8 persons have exclusively lower- or equal-status coauthors, 19 have exclusively higher-status coauthors. Thus there exists sufficient vertical as well as horizontal coauthorship and, hence, collaboration. Out of 562 coauthorship relations, 321 (*i.e.* 57.1 per cent) are within the same division.[22] While 18 persons have coauthored exclusively with persons of different divisions, 20 persons have coauthored exclusively within their divisions. Lastly, 83.4 per cent (*i.e.* 102 out of 122) of coauthorship relations are between persons of the same specialisation and 42.2 per cent (*i.e.* 76 out 166) are in the same field of activity. Thus coauthorship has greater affinity to specialisation of the space technologists than their fields of activity. Further, a sort of division of work among specialists has taken place in these cases, whereas co-operative efforts of specialists of different areas took place in other papers.

A discussion with coauthors of some selected papers revealed that less productive coauthors (*i.e.*, those with one or two papers to their credit) have not had much of useful information exchange among coauthors other than what was needed for the paper on hand. Nor they continued to exchange information after publication of the papers. However, more prolific coauthors have strongly felt the need for further coauthoring with more specialists in order to make their thoughts and papers complete. There appear to be more informal exchange of information among prolific coauthors than others, even when there is no paper being written. But, the

TABLE 8.22 : Frequency Distribution of Number of Coauthors

No. of Authors	Sample papers		
	No.	%	Cumulative %
01	10	9.6	9.6
02	8	7.7	17.3
03	8	7.7	25.0
04	1	1.0	26.0
05	8	7.7	33.7
06	21	20.2	53.9
07	11	10.6	64.5
08	10	9.6	74.1
09	12	11.5	85.6
10	4	3.8	89.4
11	3	2.9	92.3
12	1	1.0	93.3
13	1	1.0	94.3
14	2	1.9	96.2
15	4	3.8	100.0
Total	104	100.0	100.0

TABLE 8.23: Non-Intersecting Clusters of ISAC Coauthors

Group	No. of	Productivity of papers	
No.	Members	Full credit	Fractional credit
01	72	240	73.1
02	6	6	0.6
03	5	5	1.0
04	8	8	0.8
05	2	1	1.0
06	3	6	1.6
07	2	2	0.6
08	5	18	10.1
09	1	1	0.1
Total	104	287	88.9

Note : If A,B,C and D have published a paper, each one will get one paper to their credit 'ful on credit method' whereas each one will get 0.25 credit in 'fractional credit method.

focused discussion and flow of information on the specific topic was naturally maximum among coauthors while writing papers.

The reaction of respondents to the question why they coauthor a paper is very interesting. One author says that two minds are always better than one and hence there appears to be a sort of division of intellectual labour in the task. The second reason put forth was that in an experimental work assistance from others is needed and all of them cannot be just acknowledged in the paper. Thirdly, in a more inevitable situation where a meet of different minds is really needed, contributions from different areas are necessary. The exceptions to the need for coauthoring is well brought out by two of the respondents when they said that all coauthored papers are not of collaborative works or research, except where coauthors are of comparable status.

NOTES

1. Raitt (1984, p. 167) has found an almost similar result about the number of persons with whom his respondents communicated.
2. Guiding team and sharing information with the team ranked seventh as a purpose of seeking information in chapter 4.
3. This is close to the findings of Collins (1974) and Wolek (1984, p. 226).
4. The correlation coeffecients *(r)* of degree of information-sharing with the degree of dependence as informal source of information for subordinates (juniors), peers (colleagues) and superiors (seniors) are respectively, 0.33, 0.31 and 0.15.
5. The correlation coefficients *(r)* of the degree to which information is sought for the purpose of guiding team and sharing it with members of the team', with the degree of information-sharing by subordinates, peers and superiors respectively, are 0.32, 0.17 and 0.07.
6. Interestingly, those who consulted colleagues and fellow professionals for bibliographic references (Chapter 6) tend to feel more free inter-personal information-sharing by their peers ($x^2 = 8.0$, $df = 3$, $p < 0.05$).
7. Generic-type responses like 'division/section/project staff' are retained as it is in the analysis. But generic-type responses like 'project managers', 'library staff' are invalidated. Some have expressed their inability to provide five most-often contacted persons for information in the following words and they are also excluded from the analysis : 'Perons are contacted depending on the problem and situation', 'many', 'not interested to reveal', 'no regular contact for information, but will meet persons regarding work related problems', 'find it difficult to name', etc. Very few names of individuals outside ISAC have figured in more than one respondent's list of names and hence further analysis of all outside ISAC individuals is restricted to the names of the organisations to which they belong. See Appendix 6 for list of organisations.
8. A slightly higher percentage of external comunication found by Raitt is attributable to the fact that his sample consisted of fairly higher-level scientists and engineers.
9. Women space technologists do not differ from men space technologists in intra-($x^2 = 0.8$ $df = 2$, $p > 0.05$) and inter- organisational communication ($x^2 = 4.89$, $df = 2$, 0.10

> $p > .0.05$) (Sridhar, 1987).

10. For background details of concepts of relations, any fundamental book on set theory or the last part of a paper by Sridhar (1980) be seen.
11. Section is a narrower concept than division in the organisation structure *i.e.*, a division has one or more sections within it.
12. 'Linking Pin' concept is originally propounded by Lickert (1959, p. 194).
13. The information-potential of men and women space technologits respectively are 3.03 and 1.69. Nearly 80 per cent of women in the population are not frequently contacted for information by any respondent (Sridhar, 1987).
14. Bethell (1972, p. 63) found that age is not a significant point of difference but the length of service of a communication star is.
15. This is in conformity with the findings of earlier research that performance and communication are positively related (Rothwell and Robertson, 1975, p. 393; Bethell, 1972, p. 93; Pelz and Andrews, 1976, p. 47).
16. In this analysis wherever the characteristics of either person cited/contacted or person citing/contacting is not known such relations are excluded.
17. Technocrat is one who has managerial and supervisory nature of work in addition to technical nature of work and technocrats are identified by their functional designations.
18. The number of papers published ($r = 0.55$) and the number of reports prepared ($r = 0.68$) have significantly contributed to the professional activities and achievements index as they moderately and positively correlated with it.
19. The ratio of productivity of papers based on fractional credit to productivity of papers based on full credit or the ratio of total number of papers to total authorships is defined as 'productivity ratio'.
20. The only change noticed in the revised sample is that instead of three authored papers, two authored papers occur most frequently and this sample is used for recording the data about publishing, citing and coauthorship of the space technologists in the data base.
21. The frequency of multiauthored papers for past 7 years as per ISAC annual report shown in Appendix IX indicates slight decrease in the degree of collaboration and multiplicity index over years.
22. Fifty-eight per cent (294 out of 507) of coauthorship relations are within the same sections.

9

USE OF LIBRARY DOCUMENTS BY THE INDIAN SPACE TECHNOLOGISTS AND THEIR INTERACTIONS WITH THE LIBRARY

9.1 The Background about the Data on Use of Library Documents and User-interactions with the Library

The study of use and user-interactions with the 'primary library' throws further light on the information-seeking behaviour of the space technologists. The quantitative data, at the broad level, about various interactions of the IST with ISAC Library and use of library documents are presented in Table 9.1. The data, as explained below, are gathered partly based on observations for a sample period/sample collection in a preplanned way and partly extracted from the records available in the library.[1]

(i) Procurement service availed by the IST *i.e.* number of books and reports procured against request for the IST during 1982;

(ii) The number of new books, reports and journals suggested by the IST during 1982 representing the current demand for new documents and as a measure of user participation in collection development;

(iii) The number of reservations made for books and reports at the circulation counter of the library by the IST during 1982;

(iv) The number of literature searches made by the library against request on behalf of the IST (July 82—June 83);

TABLE 9.1 : Frequency Distribution of Use of Library Documents and User-interactions with the Library

Frequency	Procurement services	Suggestions for new documents	Document reservations	Literature search service	Reference service	Inter library loan service	Reprographic service(in oo's)	In house use-library visits	Borrowed use as per 'Circulation sample'	Borrowed use as per 'Collection sample'
00	462	463	336	726	606	699	449*	369	248	98
01	52	74	86	4	80	22	93	124	54	82
02	37	26	72	3	29	8	35	67	48	86
03	11	17	34	1	9	3	20	50	64	72
04	12	14	20	0	4	0	12	28	39	51
05	7	4	21	0	1	1	8	23	33	49
06	11	3	15	0	0	0	1	29	35	55
07	7	3	14	0	1	0	3	24	23	25
08	3	2	13	0	0	1	3	10	24	34
09	2	3	5	0	0	0	4	13	22	21
10	2	3	7	0	1	0	3	7	31	17
11	1	4	4	0	0	0	3	8	18	20
12	4	4	4	0	0	0	1	10	24	12
13	2	2	3	0	0	0	0	3	14	18
14	1	1	1	0	0	0	0	0	22	10
15	3	4	1	0	0	0	3	1	11	6

(Contd.)

TABLE 9.1 : Frequency Distribution of Use of Library Documents and User-interactions with the Library

Frequency	Procurement services	Suggestions for new documents	Document reservations	Literature search service	Reference service	Inter library loan service	Reprographic service (in oo's)	Inhouse use-library visits	Borrowed use as per 'Circulation sample'	Borrowed use as per 'Collection sample'
16—20	2	3	9	0	0	0	4	4	45	24
21—25	2	0	2	0	0	0	2	1	19	11
26—30	2	0	2	0	0	0	2	0	7	4
31—35	1	0	0	0	0	0	2	1	2	2
36—40	1	0	1	0	0	0	0	0	0	3
41—50	2	0	1	0	0	0	1	0	1	0
51—75	1	0	0	0	0	0	1	1	0	0
76—100	0	0	0	0	0	0	0	0	0	0
>100	0	0	0	0	0	0	0	0	1	1
Not Applicable	177	176	146	75	73	72	157	33	22	106
Error	+ 2	+ 1	0	– 2	– 3	+ 1	0	+ 1	0	—
Total	807	807	807	807	807	807	807	807	807	807

Key : *, One hundred and eleven users who consumed reprographic service upto 49 pages are treated as zero.

(v) The number of short range reference queries answered and the number of times reader assistance was provided for difficult-to-locate type of documents to the IST during July 82 and June 83.

(vi) The number of documents borrowed by the IST through inter-library loan service from other libraries during July 82 and June 83 excluding the direct borrowing facility availed with the two major libraries in the city (*i.e.* IISc and NAL libraries);

(vii) The reprographic service of the library availed by the IST in terms of number of pages of xerox copies of library-reading materials provided during 1982;

(viii) The number of visits by the IST to the library reading hall and the seat occupancy over three months (16 March 1983—17 May 1983), recorded randomly at every two hours by giving due representation to different timings of a day and different days of a week as an indirect measure of inhouse use of library documents;

(ix) Borrowed/circulation use of books, reports and journals by the IST based on three-months (July- September, 1984)'circulation sample'; and

(x) Borrowed/ circulation use of books and reports over ten months (January/March-October/December, 1983) based on a 20 per cent 'collection sample' and borrowed as well as inhouse use of two latest issues of each journal over three months (September/ December 83—January/March 84).

Certain limitations and background of the data in Table 9.1 need to be noted. The library has provided literature search and reprographic services to a limited number of users based on approval of authorised persons of the group. Though more of higher-status persons have availed of these services, everyone has access to such services with the appoval of concerned. The procurement service and suggestions for new documents are two stages of the same interaction.[2] The only difference being that the procurement service emphasises the successful meeting of demands of the users and the suggestions for new documents do not speak of whether or not of such suggestions/requirements have been fulfilled. The frequency distribution of borrowed use of library documents as per 'circulation sample' and 'collection sample' are close to each other which confirms the representativeness and the adequacy of the sample.

It is clear from Table 9.1 that the number of the space technologists who did not use/interact with library at all (*i.e.*, zero scored persons) have ranged from 98 to 726 depending on the type of use and user-interaction. While the borrowed use of documents is not made by about 30 per cent of the space technologists, document reservation was not done by 50.8 per cent, and 47.7 per cent have not figured in the inhouse use and seat occupancy data. Naturally, as many as 95.1 per cent have not used inter-library loan service, about 82 per cent have not availed of the reference service and 73 per cent have not suggested any documents and hence not directly availed of any procurement service. Further, except one or two, none of the users have availed of reference service and inter-library loan service more than five times. In case of borrowed use, inhouse use, reprographic service, suggestions and procurement services intensive interactions upto 50 times by few users are observed. Keeping in view the general drawback of user-studies that they ignore nonusers, views of the nonusers are elicited through questionnaire, interview and discussions. The major reasons, for not using library are: (i) Lack of time due to the time-bound nature of projects; (ii) The work does not demand use of literature/documentary information on a regular basis, *i.e.* they rush to library either when they have to prepare reports or papers or prepare for departmental review for promotion or if they are undergoing part-time studies; (iii) The culture in general and the organisation culture in particular are not in favour of use of formal documentary information and therefore of the library; and (iv) The attitude of an individual. For instance, one senior engineer frankly admits that he did not use libraries even for his postgraduate level studies.

9.2 Library-Use Index and Library-interaction Index

In order to simplify the analysis of various types of use of library documents and various user-interactions with the library, two indexes have been developed based on individual frequencies[3] depicted in Table 9.1. The first index called the 'Library-Use Index' is defined as the sum of: (i) one-hundredth of total number of pages of xerox copies of library reading materials taken during a year; (ii) two times the number of documents borrowed through inter-library loan during a year; (iii) five times the number of visits paid to the library and reading seat occupied during the three months sample survey; and (v) the mean borrowed/circulation use of books, reports and journals of the circulation and the collection samples. The second index called 'Library-Interaction Index' is defined as the sum of: (i) the number of books and reports procured for a user; (ii) the number of new books, reports and journals suggested by a user; (iii) the number of reservations made for books and reports by a user; (iv) two times the number of literature searches provided to a user, and (v) the number of short range reference queries answered and reader assistance service availed by a user during one year duration.

Table 9.2 depicts the frequency distribution of 'library-use index' and 'library-interaction index' values for the population. It is clear from the table that as many as about 40 per cent of the population have no 'interactions' with the library and hence scored zero under 'library-interaction index'. However, the number of users who scored zero under 'library-use index', (*i.e.* the non-users of library documents) are just about 12 per cent. Further, nearly 28 per cent of the users have scored more than the mean value of 'library-use index' (19) as well as 'library-interaction index' (6). Both the distributions are quite skewed.

9.3 Correlation of Library Use Index and Library Interaction Index with User-characteristics

The results of statistical association/correlation tests of the six user-characteristics with 'library-use index' values and 'library-interaction index' values are shown in Table 9.3. One can see an almost perfect positive rank order correlation ($r_s = 0.99$) of status of the users (which supports the findings of Shaw, 1956, p. 17, 20, 48-49) and a moderate positive correlation ($r = 0.4$) of professional activities and achievements with both the indexes (which supports the findings of Lufkin and Miller, 1966, p.180 and Case Institute of Technology, 1960, p.21). The qualifications, nature of work and specialisation of the users also exhibited statistically significant relation with both the indexes. However, experience has shown a slight negative correlation ($r = -0.12$) with library use index supporting the findings of the Bath University Library (1971), Barkey (1966) and Ford (1977, p. 93) and contrary to that found by Scott (1960, p.29), Lipetz (1970), Fearn and Melton (1969). There is no significant correlation of experience with library interaction index.[4]

The mean score of 'library-use index' and 'library- interaction index' against status, qualifications, nature of work and specialisation of the users are shown, respectively, in Tables 9.4, 9.5, 9.6, and 9.7. As is evident from Table 9.4, persons in the top ranked status made 27 times more use of library documents and 132 times more interactions with the library than lowest status persons, confirming thereby the under-previleged nature of lower status technicians (Slater and Fisher, 1969, p 17-18 and 50). Table 9.5 reveals that the qualifications of the users linearly and positively

TABLE 9.2 : Frequency Distribution of Library-use Index and Library-interaction Index

Frequency	Library-use Index			Library-interaction Index		
	No. of users	% of users	Cumulative % of users	No. of users	% of users	Cumulative % of Users
00	75	11.9	11.9	253	40.2	40.2
01	51	8.1	20.1	64	10.2	50.4
02	38	6.1	26.1	70	11.1	61.5
03	25	4.0	30.1	46	7.3	68.8
04	27	4.3	34.4	30	4.8	73.6
05	25	4.0	38.4	18	2.9	76.5
06	24	3.8	42.2	13	2.1	78.6
07	21	3.3	45.5	13	2.1	80.7
08	20	3.2	48.7	15	2.4	83.1
09	21	3.3	52.0	11	1.7	84.8
10	13	2.1	54.1	13	2.1	86.9
11–15	58	9.2	63.3	24	3.8	90.7
16–20	60	9.6	72.9	23	3.7	94.4
21–25	35	5.6	78.5	8	1.3	95.7
26–30	24	3.8	82.3	6	1.0	96.7
31–40	43	6.8	89.1	9	1.4	98.1

(Contd.)

TABLE 9.2 *(Contd.)*

Frequency	Library-Use Index			Library-Interaction Index		
	No. of users	% of users	Cumulative % of users	No. of users	% of users	Cumulative % of Users
41–50	22	3.5	92.6	7	1.1	99.2
51–75	37	5.9	98.5	5	0.8	100.0
76–100	7	1.1	99.6	1	0.2	100.2
100–150	2	0.3	99.9	0	0.0	100.2
Total	628	99.9	99.9	629	100.2	100.2

Notes : 1. Some components of the indexes were not available/applicable in case of 179 and 178 respectively for library-use index and library-interaction index and hence all those cases have been excluded from the table.

2. The mean values of library-use index and library-interaction index for population as a whole are respectively 19 & 6.

TABLE 9.3 : Results of Association/Correlation Tests of Library-use Index and Library-interaction Index with User-characteristics

User-characteristics	Library-use Index			Library-interaction Index		
	$r_s/x^2/r$	*df*	Tabulated value of x^2 at $p = 0.05$	$r_s/x^2/r$	*df*	Tabulated value of x^2 at $p = 0.05$
Status (r_s)	0.99*	8	—	0.99*	8	—
Qualifications (x^2)	225.27*	16	26.296	146.86*	8	15.507
Nature of work (x^2)	84.67*	24	36.42	39.50*	12	21.026
Specialisation (x^2)	75.92*	20	31.41	30.53*	10	18.31
Experience (r)	–0.12*	290	—	– 0.004	292	—
Professional Activities and Achievements Index (r)	0.38*	174	—	0.39*	178	—

Key : *, The association/correlation is statistically significant at 0.05 significance level.

related to 'library-use index' and 'library-interaction index'. Doctorates being intensive users of the library, used the library nearly seven times that of an average undergraduate, three and a half times that of an average

TABLE 9.4 : Status Vs Library-use Index and Library-interaction Index

Status	Library-use Index (Mean)	Library-interaction Index (Mean)
A,B,C & D	26.12	15.85
E	22.22	8.63
F	22.84	7.38
G	21.09	5.95
H	17.16	3.96
I	14.92	3.93
J	9.60	1.21
K	3.86	0.27
L	2.80	0.68
M & N	0.96	0.12
Total	18.80	5.90

($r^S = 0.9879$, $df = 8$, $p < 0.01$) ($r^S = 0.9879$, $df = 8$, $p < 0.001$)

Note : Status A to N are in the descending order of hierarchical rank.

TABLE 9.5 : Qualifications Vs Library-use Index and Library-interaction Index

	Qualifications	Library-use Index (Mean)	Library-interaction Index (Mean)
U	Undergraduate	4.9	0.9
D	Diploma Holder	9.7	2.3
B	Graduate	22.9	8.4
M	Postgraduate	29.5	7.8
P	Doctorate	33.5	21.0
	Total	19.0	6.1

($x^2 = 225.27$, $df = 16$, $p < 0.05$) ($x^2 = 146.86$, $df - 8$ $p < 0.05$)

Note : The actual contingency tables on which chi-square test is made are not shown. The ranges used in the contingency table for library-use index are 0—2, 3—10, 11—20, 21—30 and more than 30. In case of library-interaction index the ranges are 0—2, 3—10, and more than 10.

diploma holder, one and a half times that of an average graduate and 1.14 times that of an average postgraduate in the population. As far as the number of interactions with the library are concerned, an average doctorate interacted more than 23 times that of an average undergraduate, nine times that of an average diploma holder, two and a half times that of an average graduate and 2.7 times that of an average postgraduate. This very much supports the findings of Scott (1960, p.16), Lipetz (1970), Fearn and Melton (1969) and Shuchman (1981, p.35). As it could be seen from Table 9.6 those who did planning and system analysis work have made maximum use of the library and interacted to the maximum with the library, followed by design and development personnel and managers and supervisors than others (who have made less than average use of and interactions with the library). This contradicts the finding that managers and supervisors make maximum use of library (Scott, 1960, p.28, Shuchman, 1981, p.36 and Raitt, 1984, p.278). From Table 9.7 one can see that physicists, mathematicians and aeronautical and structural engineers have made more use of and interactions with the library than all others (who have made less than average use of and interactions with the library) Though this supports the general finding of the earlier studies that scientists make more use of library than engineers, the aeronautical and structural engineers in the present study are exceptions.

Table 9.6 : Nature of Work Vs Library-use Index and Library-interaction Index

Nature of work		Library-use Index (Mean)	Library-interaction Index (Mean)
A	Management/ Supervision	21.5	8.6
B	Planning/ System analysis	30.8	9.0
C	Design and development	24.6	8.4
P	Design, development, Fabrication and Testing (C & D)	14.9	4.1
D	Fabrication and testing	11.0	3.0
E	Operational activity	12.5	2.1
F	Others	9.6	4.3
	Total	18.9	5.9

($x^2 = 84.67$, $df = 24$, $p < 0.05$) ($x^2 = 39.50$, $df = 12$, $p < 0.05$)

Note: The contingency tables on which chi-square test is made are not shown. The ranges used in the contingency table for library-use index are 0—2, 3—10, 11—20, 21—30 and more than 30. In case of library-interaction index the ranges are 0—2, 3—10, and more than 10.

9.4 Use of Library Documents

The responses to the question "To what extent the information has helped you in your work?" varied widely in the interviews. The theoretical and basic science oriented persons have felt to a great extent the impact of information on their work, and by and large, they considered papers as their 'backbones'. In case of developmental personnel, the information is of a limited help in their work. To quote one, "The information from written documents is not often directly applied in the field. Formal documents help us to have intuitive guess about future work". It is interesting to note that Collins (1974, p.176), in his study of scientists working on TEA lasers, called published documents as 'Pseudo Publications' as the early publications on TEA lasers were misleading than helpful and kept back certain amount of information. In fact, a similar reaction is found from one of the respondents about the significance of information in technical reports. He felt that normally some link would be missing in

TABLE 9.7 : Specialisation Vs Library-use Index and Library-interaction Index

Specialisation		Library-use Index (Mean)	Library-interaction Index (Mean)
A	Physics	34.5	12.7
C	Mathematics	31.8	8.4
E	Mechanical engineering	10.7	2.8
F	Aeronautical and structural Engineering	29.0	10.0
H	Electrical engineering	15.1	3.8
I	Electronics	17.8	6.0
	Total	18.0	5.8
		($x^2 = 75.92$, $df = 20$, $p < 0.05$)	$x^2 - 30.53$, $df - 10$, $p < 0.05$)

Note : The contingency tables on which chi-square test is made are not shown. The ranges used in contigency table for library-use index are 0–2, 3–10, 11–20, 21–30 and more than 30. In case of library-interaction index the ranges are 0–2, 3–10 and more than 10.

the information provided in technical reports and this often made it difficult to repeat the work reported.

It is felt by the respondents that the results of a work is proportional to or is commensurating with information collected and used upto a certain basic level and, thereafter, though the information is valuable and crucial, proportional benefits cannot be expected. In many situations, the information is not immediately put into use. This is felt so because the space technologists cannot always go for innovative plans; they depend heavily on past experience, partly due to the high reliability needed and partly due to the cost and risk involved. It is necessary to strike a balance among above pragmatic factors while doing innovation based on new information.

However, It is revealed during the interview that the information collected has helped the space technologists in preparing reports and new proposals, knowing latest developments, formulating and solving specific problems, making the end results more accurate, filling blanks in the development and research in narrow fields like TDPs, improving an existing method/system, etc. Finally, it is also felt that irrespective of whether information has helped them in their present work or not, they have to acquaint themselves with the latest developments in order to keep themselves up-to-date. This observation is close to that found by Raitt (1984, p.236) that "... the average scientist or engineer uses the library more to see what new books and journal issues are available rather than as a real source of information".

During the interview, it was also discussed with the respondents whether they used documents sequentially (consecutively) or simultaneously (conjunctively). Nearly 60 per cent of the respondents said that most of the times they used documents consecutively and the rest conjunctively. The opinion was equally divided on this issue in INFROSS study (Line, 1971, p.422). The need for conjunctive reading/consultation of documents arose for specialised task-oriented use in a specific area, when a generalist wants to review the literature to gain an indepth understanding on a specific topic and when one writes a paper necessitating quoting from the various documents. Lastly, there appears to be a need to orient engineers and other technical staff to use of information and library.

9.4.1 *Circulation (or Borrowed or Lentout) Use of ISAC Library Documents*

The lentout use study (Sridhar, 1985, p.26) of ISAC Library has shown some clear trends in the use of books, reports, journals and standards by its users. Books are used intensively, journals moderately, reports marginally and standards negligibly at ISAC Library. Based on a 20 per cent 'collection sample', it is found that books and reports are lentout, on an average, 1.51 and 0.22 times respectively during the ten months. In case of current journals, the average number of times a current issue is lentout is 1.22 in a quarter (Sridhar, 1986, p.77—80). The analysis of a three months 'circulation sample' of lentout use of ISAC Library documents revealed that books consitituted 80 per cent of daily issues followed by journals with 14.8 per cent, reports with a meagre 5 per cent and the standards with a negligible 0.2 per cent. Like inhouse use, lentout/borrowed use of library documents has also increased during departmental reviews for promotion and on a typical day it followed the bimodal roughly symmetric distribution (Sridhar, 1982). But over a typical week, the borrowed use is in the reverse pattern of inhouse use *i.e.*, minimum during middle of the week and maximum during beginning and end of the week. During 3 months study of circulation sample, 69 per cent of members of the ISAC library borrowed one or more documents (*i.e.*, the ratio of potential users to total membership) which is significantly higher than that found in earlier studies (Slater and Fisher, 1969, p.21; Scott, 1960,

p. 3; Bush *et al*, 1956, p.94 and Shuchman, 1981, p.30).

9.4.2 *Use of Books of ISAC Library*

Use of books, based on 20 per cent stratified 'collection sample', showed that nearly 27 per cent of the books are never borrowed and 21 per cent of the books accounted for 83 per cent of the total use following typical 80/20 rule. In a longitudinal analysis certain subject-groups such as electronics and manufactures have shown an increase in percentage of use in recent months. Use did not correlate clearly with age (year of publication) of books. Yet the average use per book increased more uniformly, steadily and steeply in case of use over year of acquisition than what it does in the case of use distribution over age of book (Sridhar, 1986).

As far as the defined population of this study is concerned, it is found that 29.2 per cent of the space technologists did not borrow (use) any books during sample (collection) period. When the data of the use of books is examined against broad subjects as per UDC scheme for classification, it is found that among those who borrowed books, 38.8 per cent have borrowed books from one subject, 30.4 per cent borrowed from two subjects, 18.8 per cent borrowed from three subjects, 6.8 per cent borrowed from four subjects and the rest borrowed books from 5 to 8 subjects (except one who borrowed from 12 subjects). In other words, among those who used books, little over two-third are satisfied with book collection in one or two subjects. Generally, one who borrowed from more subject groups tended to borrow more books. In other words, the minority of less than one-third of the actual borrowers have not only made heavy use of books but also needed books from more than two subjects to satisfy their multidisciplinary and interdisciplinary needs.[5]

9.4.3 *Use of Reports of ISAC Library*

Use of 20 per cent systematic random sample reports showed scarce or marginal use (Sridhar, 1984). Nearly 78 per cent of sample reports are not issued out even once and less than 8 per cent and 12 per cent of the sample accounted respectively for 67 per cent and 85 per cent of the total circulation use of the reports. Thus the frequency distribution of reports is more skewed than that of books. Technical reports are believed to be very popular among space scientists and technologists throughout the world; these are considered to be one of the principal media for primary communication bestowed with several advantages such as user directed nature, ease, speed and flexibility in presentation and production (Sridhar, 1982). Yet the Indian space technologists have underused the reports at their primary library. The reasons for limited use of reports from the library collection are: reports collection of the library is not strong, having been developed only since 1978; reports are available to users directly as gratis from originating sources and authors; the internal/Indian reports of the library are otherwise accessible to many users; fairly adequate and free reprographic service has enabled many users to possess their own copies of relevant pages or complete report itself; and lastly reports normally deal with highly specific and narrow topic and attract only a limited number of users working on the topic. NASA reports forming 27 per cent of the sample have accounted for 53 per cent of use and ESA reports forming 9 per cent of the sample accounted for 15 per cent of use. Reports of foreign origin forming 85 per cent of the sample accounted for 92 per cent of total use whereas 15 per cent of Indian reports resulted in 8 per cent of total use. In a longitudinal analysis, it was found that use of foreign reports in general and reports of NASA, ESA, RAE, RAC, NTIS in particular, have shown increased use during the recent months. The reports collection has about 97 per cent of the reports brought out during the last 12 years and the age of reports and the year of acquisition of reports did not show any significant relation to use. Though the space technologists have indicated a cutoff time of 20 years about utility of old documents and information, there is no linear decrease in use of older documents of less than 20 years.[6]

9.4.4 *Use of Journals of ISAC Library*

A use study based on borrowed as well as inhouse use of two latest issues of 485 current journals of ISAC Library (Sridhar, 1986) showed a moderate use with 4.71 and 1.22 average inhouse use and borrowed use respectively per issue per quarter. The borrowed use per person per quarter per single issue of all journals is 1.6. It is also found that about 12 per cent of journals are not used at all. About 20 per cent of the journals accounted for nearly 52 per cent of the total use of journals. Thus frequency distribution of use of journals is much less skewed than that of books and reports. In other words, the interest of the space technologists is more widely spread among different journals than in case of books and more widely among books than in case of reports. Inhouse use of the current journals (the number of issues as well as titles) is much more than lentout use as 43.5 per cent of the journals are never borrowed. This supports the findings of the citation study (Sridhar, 1985) that nearly 60 per cent of the journal articles cited by the space technologists are from ten journals. Further, a slight positive correlation ($r = 0.19$) between publishing in and citing from a journal is noticed. But use of a journal is found to be independent of publishing articles in it and citing from it by the users (Sridhar, 1986).

The abstracting and indexing journals are relatively less used confirming the finding of the survey through questionnaire. Journals in the core areas are used to the maximum. The average number of uses per title (journal) per quarter per two issues revealed that science, physics, electronics and aeronautics, astronautics and spacefight journals topped the list. The lowest average use was recorded by abstracting and indexing journals followed by energy and power systems, mathematics and statistics and quality and reliability engineering journals.

Only 31.9 per cent of library members borrowed journals during three months and 38 per cent of those borrowed have made 74 per cent of total borrowings of journals. More than half of the users borrowed issues of a single journal/title in the area of their interest, nearly 22 per cent of the users borrowed issues of two journals/titles, 10 per cent borrowed issues of three journals/titles, 15.5 per cent borrowed issues of 4 to 10 journals/titles and rest of the 6 users borrowed 11 or more journals/titles. In other words, nearly three-fourth of the users have borrowed journal issues of one or two titles only. Though these figures are based on borrowed use of journals, the number of journal titles used by an average space technologist is much lower than 5-15 journals read by an average scientists as per previous studies (Scott, 1959, p. 113; Martyn, 1964, p. 20; Bernal, 1948; Martin, 1962; Wood and Hamilton, 1967; Gralewska-Vickery, 1976, p. 274; Hanson, 1964, p. 65 and Ford, 1977, p. 32). However, by adding personal subscriptions and inhouse use of journals and deleting lower-level technicians from the present population, one can hope to find a comparable figure in case of the IST. As far as the defined population of the space technologists is concerned, little over two-third (67.6%) of the users have not borrowed the sample issues of journals during a quarter. (Raitt 1984, p. 227) also found lower use of journals). Among those who borrowed, 54 per cent borrowed journals from one subject, 26.8 per cent borrowed from two subjects, 8 per cent borrowed from three subjects, 6.9 per cent borrowed from four subjects and the rest borrowed journals from 5 to 10 subjects. In other words, among those who used journals, more than 80 per cent of them are satisfied with journals relating to one or two subjects only.[7]

9.4.5 *Use of Other Types of Documents of ISAC Library*

Among the other types of documents in the library, bound journals are not lent out much and standards are found to be negligibly borrowed (Sridhar, 1984). Except ISI, no other standards are ever borrowed. Out of 630 sample standards, ISI standards have alone been issued 15 times during a ten month period. Like reports, the collection development of standards was also started from 1978 and standards other than ISI are acquired much recently. Secondly, many standards such as MIL and JSS are decentralised and concerned groups in ISRO have much stronger collection of these standards than the library. The library's standards collection was enriched in recent years by partially transferring the collection from such groups and by receiving as gratis from individuals. The lending policy for standards (*i.e.* 3 days without renewing facility) and the requirement of occasional consultation of standards made users prefer inhouse use of standards than borrowing.

Even the inhouse use of microfiche reports is limited and borrowed use is negligible. Though the library has about 800 trade catalogues, their use is also limited as majority of them are free catalogues and the concerned divisions, sections, projects as well as individual users have pertinent collection of trade catalogues with them. Hence the use of such documents has not been examined in greater detail.

9.5 Correlation of Use of Library Documents with User-characteristics

The results of correlation/association tests of borrowed use of library documents with user characteristics is shown in Table 9.8. Tables 9.9, 9.10, 9.11, and 9.12 respectively present the status, qualifications, nature of work and specialisation versus the mean borrowed use of books, reports and journals together with the average number of journal titles used.[8] As could be seen from the Tables 9.8 and 9.9 there exists a statistically significant moderate positive rank order correlation of borrowed use of books ($r_s = 0.76$) and borrowed total use of books, reports and journals ($r_s = 0.68$) with the status of the users. However, borrowed use of reports and journals exhibited a very high positive rank order correlation ($r_s = 0.92$) with status of the users. In other words, reports and journals are used mostly by higher level personnel.

Tables 9.8 and 9.10 clearly shows statistically significant association of borrowed use of books, reports and journals with qualifications of the

TABLE 9.8 : Results of Association/Correlation Tests of Borrowed use of Library Documents with User-characteristics

User-Characteristics	Books (For 10 months)		Reports (For 10 months)		Journals (For 3 months per 2 issues of 485 titles)		Total use @		Av. no. of journals / titles used	
	$r_s/x^2/r$	*(df)*	$r_s/x^2/r$	*(df)*	$r_s/x^2/r$	*(df)*	$r_s/x^2/r$	*(df)*	$r_s/x^2/r$	*(df)*
Status (r_s)	0.76*	(8)	0.92*	(8)	0.93*	(8)	0.69*	(8)	0.89*	(8)
Qualifications (x^2)	83.18*	(12)	45.37*	(4)	85.80*	(12)	125.83*	(16)	82.64*	(8)
Nature of work (x^2)	25.62	(18)	12.75*	(6)	40.35*	(18)	52.31*	(24)	31.02*	(12)
Specialisation (x^2)	30.29*	(15)	32.29*	(5)	27.92*	(15)	48.44*	(20)	23.07*	(10)
Experience	-0.10	(377)	0.08	(396)	-0.04	(392)	-0.08	(390)	-0.04	(390)
Professional activities and achievements Index *(r)*	0.11	(169)	0.27*	(179)	0.08	(177)	0.15*	(176)	0.06	(176)

Key : *, The association/correlation is statistically significant at $p = 0.05$; @, total is not the total of the preceding columns but mean of the total use of documents.

users. The relation is positive and almost linear except that postgraduates tend to use more books and journals than doctorates. This very much supports Shuchman's (1981, p. 35) finding that "resource use of library, librarian, and data bases increases moderately with education". The use of reports and journals is almost confined to graduates, postgraduates and doctorates. Interestingly, postgraduates have used more books than doctorates (and others), and further, postgraduates and graduates have used more journals than doctorates (and others). But use of reports by doctorates is more than double that of postgraduates and five times that of graduates.

As previously found in the analysis of 'library-use index', one can see from Tables 9.8 and 9.11 that planning and system analysis personnel have made maximum borrowed use of library documents than others. The use of reports, journals and total use of documents (but not books) have shown statistically significant relation with the nature of work. The relation is not strictly linear. By and large, borrowed use of reports and journals is confined to management/supervision, planning/system analysis, design and development personnel. Shuchman's (1981, p.36) finding

TABLE 9.9 : Status Vs Mean Borrowed use of Library Documents (Collection Sample)

Status	Books	Reports	Journals	Total@	Journals/ Titles
A,B,C & D	4.46*	0.29	4.50*	4.64	1.26
E	3.24	0.29	1.69	5.22	1.19
F	4.57	0.54	1.30	6.31	1.06
G	4.71	0.18	1.86	6.46	1.31
H	4.52	0.09	1.40	6.00	1.09
I	4.28	0.08	1.09	5.46	0.80
J	2.75	0.02	0.75	3.52	0.66
K	1.76	0.03	0.16	1.95	0.16
L	1.48	0.00	0.23	1.71	0.23
M & N	1.16	0.02	0.13	1.29	0.13
Total	3.95	0.16	1.26	4.81	0.70

Key : *, Includes one high executive to whom documents were issued on standing instructions; @, total is not the total of the preceding columns but mean of the total use of documents.

Note : Status A to N are in the descending order of hierarchical rank.

that 'the library is a source of information more often for those in nonmanagerial positions than for those with managerial responsibility' is not true in the present study. So is the case about the finding of Raitt (1984, p.278) that ". . . the library or information centre has not been regarded as essential or even useful by many managers . . . and the information contained in the library is usually not the kind that managers want".

As could be seen from Tables 9.8 and 9.12 there appears to be a statistically significant relation of specialisation of the space technologists with use of books, reports and journals. Not surprisingly, the mathematicians have made maximum use of books (almost double the average use of others) and journals and aeronautical and structural engineers have made more than double the use of reports than others.

As is evident from Table 9.8 experience and professional activities and achievements of the space technologists have no influence on the use of books, reports and journals except a low positive correlation ($r = 0.27$) of professional activities and achievements with the use of reports. The findings of the earlier studies (Ford, 1977, p. 32 and Scott, 1960, p. 28) that the number of journals read by users increased with the age[9] and experience is not supported by the present study. Among other things, Raitt (1984, p. 278-279) attributed low usage of library to the fact that his

TABLE 9.10 : Qualifications Vs Mean Borrowed use of Library Documents (Collection Sample)

Qualifications		Books	Reports	Journals	Total use @	Journals/ Titles
U	Undergraduate	1.68	0.06	0.25	2.19	0.21
D	Diploma holder	2.79	0.04	0.58	3.62	0.50
B	Graduate	3.63	0.14	1.22	5.76	1.06
M	Postgraduate	4.01	0.27	1.18	6.20	0.95
P	Doctorate	3.60	0.68	0.87	5.61	0.77
	Total	3.16	0.16	0.87	4.70	0.73

Key : @, Total is not the total of the three preceding columns but mean of the total use.

TABLE 9.11 : Nature of Work Vs Mean Borrowed use of Library Documents (Collection Sample)

Nature of work		Books	Reports	Journals	Total use @	Journals/ Titles
A	Management/supervision	3.82	0.12	1.17	5.42	1.05
B	Planning/system analysis	4.58	0.42	1.47	7.47	1.15
C	Design and development	3.91	0.18	1.14	6.04	0.99
P	Design, development, Fabrication and testing (Both C & D)	3.31	0.07	0.96	4.39	0.76
D	Fabrication and testing	2.81	0.10	0.48	3.62	0.41
E	Operational activity	2.97	0.03	0.61	4.13	0.56
F	Others	2.47	0.00	1.44	4.25	1.11
	Total	3.51	0.16	0.96	5.16	0.81

Key : @, Total is not the total of the preceding columns but mean of the total use of documents.

respondents were fairly experienced (and aged) and well settled in jobs with less job turnover. On the other hand, though the IST, in the present study, are young, less experienced with high job turnover and showed strong motive to keep themselves up-to-date in their field and used the library fairly better, their age and experience do not seem to be contributing much for the use of library.

TABLE 9.12 : Specialisation Vs Mean Borrowed use of Library Documents (Collection Sample)

Specialisation	Books	Reports	Journals	Total use @	Journals/ Titles
A Physics	3.88	0.24	0.88	5.35	0.76
C Mathematics	5.55	0.25	1.35	9.15	1.15
E Mechanical Engineering	1.89	0.11	0.68	3.66	0.57
F Aeronautical and Structural Engineering	3.97	0.50	0.69	6.31	0.50
H Electrical Engineering	3.15	0.18	0.76	4.68	0.50
I Electronics	3.62	0.09	0.97	5.22	0.81
Total	3.44	0.14	0.88	4.99	0.73

Key : @, Total is not the total of the three preceding columns but mean of the total use of documents.

Lastly, the professional activities and achievements have a slight positive correlation ($r = 0.27$) with use of reports but not with use of books and journals, which do not support the finding of Scott (1960, p. 28) that professional activities and the number of journals used are correlated.

9.6 Inhouse and Inter-library Loan Use of Library Documents and Use of Documents Through Reprographic Service

9.6.1 *Reading Seat Occupancy and Inhouse use of Library Documents*

The in house use of current journals by the IST is discussed along with borrowed use of journals in the previous sections. However, the inhouse use of books, reports and other documents of ISAC Library could not be measured directly due to practical difficulties. But an attempt has been made to assess inhouse use of library documents in an indirect way by monitoring the reading seat occupancy of the IST in the library. The library visit and reading seat occupancy data were noted randomly over three months (Sridhar, 1982). The total data collected was of the order of 1719 user-visit/seat occupancy by 429 individuals (53%) as against the ratio of visits to enrolment of 37 per cent in another study (Bush, *et. al.*, 1956, p. 88). The distribution of the data against users is skewed and 17.5 per cent of the users accounted for 56.7 per cent of visits. The data revealed that the inhouse use of documents during a typical working day followed bimodal roughly symmetric distribution and over a typical week cyclical distribution with maximum in the middle of the week and minimum at the beginning and end of the week. Further, during the season of departmental reviews for promotion, there is a considerable increase in the inhouse use of library documents and the major events/activities of the centre have caused swift variations in the inhouse use of library documents. It is estimated that on an average five documents are used per visit per user. Female users have made an average of 3.92 inhouse user visit as against 2.48 by the male users.

Further, except intensive inhouse users, more inhouse use is made by the users of nearby divisions of the library than others. The rank order correlation of the average library visits for inhouse use of documents by the members of 13 divisons/projects with the 'nearness' of the division/project to the library showed a statistically significant moderate positive correlation ($r_s = 0.55$). However, the inverse square law i.e. the amount of use of a library is inversely related to the square of the distance between the library and the functional group to which a user belongs does not hold good. Interestingly, users working in projects (average is 4.09) have made more inhouse use than other functional divisions (average is 3.17) and service sector divisions (*i.e.* EMD and Facilities) have made least (average is 0.65) inhouse use of the library.

9.6.2 *Correlation of Inhouse Use of Library Documents with User-Characteristics*

Results of association/correlation tests of inhouse use, inter-library loan use and use of documents through reprographic service with user-characteristics are shown on Table 9.13. Tables 9.14, 9.15, 9.16 and 9.17 respectively provide the mean inhouse use, inter-library loan use and use of documents through reprographic service against status, qualifications, nature of work and specialisation of the space technologists. As evident from Table 9.13, the inhouse use of the library correlated highly with status ($r_s = 0.88$), related significantly to qualifications, nature of work and specialisation and slightly and positively correlated with professional activities and achievements of the space technologists (r=0.21).[10] Particularly, postgraduates followed by graduates and doctorates (Table 9.15), planning/system analysis personnel followed by design and development personnel (Table 9.16) and mathematicians followed by physicists, aeronautical and structural engineers and electronics engineers

TABLE 9.13 : Results of Association/Correlation Tests of Inhouse Use-Inter-library Loan Use and Use of Documents through Reprographic Service with User-characteristics

User-characteristics	Inhouse Use/ Seat Occupancy		Inter-library Loan Requests		Use through Reprographic Service	
	$r_s/x^2/r$	*(df)*	$r_s/x^2/r$	*(df)*	$r_s/x^2/r$	*(df)*
Status *(r_s)*	0.88*	(6)	0.90*	(6)	0.93*	(6)
Qualifications *(x^2)*	160.25*	(16)	60.01*	(8)	142.35*	(12)
Nature of work *(x^2)*	72.65*	(24)	15.70	(12)	90.70*	(18)
Specialisation *(x^2)*	53.41*	(20)	20.50*	(10)	67.98*	(15)
Experience *(r)*	–0.06	(395)	–0.02	(396)	0.07	(395)
Professional Activities and Achievements Index *(r)*	0.21*	(179)	0.24*	(179)	0.53*	(179)

Key : *, The association/correlation is statistically significant at $p = 0.05$

TABLE 9.14 : Status Vs Inhouse User, Inter-Library Loan Use and Use of Documents through Reprographic Service

Status	Average Inhouse Use/ seat occupancy	Average interlibrary loan requests	Average use through reprographic service (in 100 pages)
A,B,C & D	3.80	0.54	12.07
E	3.56	0.16	3.85
F	3.24	0.07	1.95
G	3.76	0.15	1.63
H	3.50	0.01	0.70
I	2.22	0.02	0.40
J	1.75	0.00	0.90
K, L, M, & N	0.39	0.01	0.03
Total	2.58	0.08	1.74

Note : The status A to N are in the descending order of hierarchical rank.

TABLE 9.15 : Qualifications Vs inhouse use, Inter-Library Loan Use and Use of Documents through Reprographic Service

Qualifications		Average inhouse use seat occupancy	Average inter-library loan requests	Average use through reprographic service (in 100 pages)
U	Undergraduate	0.61	0.01	0.10
D	Diploma holder	1.46	0.02	0.34
B	Graduate	3.21	0.06	1.55
M	Postgraduate	4.18	0.16	3.74
P	Doctorate	3.11	0.57	5.96
	Total	2.60	0.09	1.78

TABLE 9.16 : Nature of Work Vs Inhouse Use, Interlibrary Loan Use and Use of Documents through Reprographic Services

Nature of Work		Average inhouse use/ seat occupancy	Average interlibrary loan requests	Average use through reprographic service (in 100 pages)
A	Management/ Supervision	2.39	0.23	6.76
B	Planning system Analysis	4.23	0.11	5.65
C	Design and Development	3.96	0.12	1.55
P	Design, Development, Fabrication and Testing (Both C & D)	2.79	0.08	1.50
D	Fabrication and Testing	1.72	0.04	0.37
E	Operational Activity	1.62	0.00	0.22
F	Others	2.70	0.00	0.00
	Total	2.95	0.09	2.05

TABLE 9.17 : Specialisation Vs Inhouse use, Inter-Library Loan Use and Use of Documents through Reprographic Service

	Specialisation	Average inhouse use/ seat occupancy	Average inter-library loan requests	Average use through reprographic service (in 100 pages)
A	Physics	3.73	0.31	5.56
C	Mathematics	4.10	0.00	0.82
E	Machanical engineering	1.34	0.04	1.17
F	Aeronautical and structural engineering	3.28	0.13	8.46
H	Electrical engineering	2.26	0.00	3.12
I	Electronics	3.19	0.07	1.53
	Total	2.79	0.08	2.10

(Table 9.17) have made more inhouse use of library than others. The inhouse use of library made by managers/supervisors is ostensibly much less than that of planners/system analysts and design and development personnel.

9.6.3 *Use of Documents of other Libraries Through Inter-library Loan Facility*

The number of documents supplied to the space technologists as part of inter-library loan facility during July 1982 and June 1983 were recorded and analysed. As already mentioned, this data excluded the use of IISc and NAL libraries from which the space technologists could directly borrow by using inter-library borrower tickets. Since the inception of ISAC library in late 1972 till 1979/1980, the space technologists have heavily depended on other local libraries especially the IISc and NAL libraries with which institutional arrangements have been made to visit as well as directly borrow documents. The dependence on these libraries has been very much reduced after 1979/1980. An attempt was made during April 1983 and June 1983 to record all the documents directly borrowed from IISc and NAL libraries by the population of the study. Though the data appeared to be slightly under represented due to difficulties faced by the respective libraries and refusal of some users to record such transactions, the total borrowing of 24 documents from IISc library and 14 documents from NAL library gave many clues about the use of these two libraries by the space technologists. The 24 documents borrowed from the IISc library, comprised 17 books, 6 back volumes of journals and one technical report. In case of the NAL library, it was 12 books and 2 back volumes of journals. The data clearly confirm the general belief that the space technologists have been depending very much on these libraries for back volumes of journals but not on technical reports. Most of the books borrowed from these libraries are of general interest in nature and almost half of them are also available at ISAC library. But as they are in high demand at ISAC Library, many users have sought them from other libraries.

9.6.4 *Correlation of Use of Documents Through Inter-library Loan Service with User-characteristics*

Table 9.13 shows that the use of inter-library loan service by the IST correlated very highly and positively with status (r_s=0.88), slightly and positively with professional activities and achievements (r=0.24), related significantly to the qualifications, and specialisation of users. Particularly, the inter-library loan service is availed mostly by persons of status G and above (Table 9.14), doctorates (Table 9.15), and physicists followed by aeronautical and structural engineers (Table 9.17) than others. The use of inter-library loan service by an average doctorate is more than three times that of an average postgraduate and more than nine times that of an average graduate.

9.6.5 *Use of Library Documents Through Reprographic Service*

The ISAC library provided reprographic service to the tune of 34,000 pages of xerox copies of reading materials from the library in addition to over 53,000 pages of xerox copies of reprints, content pages of new journals, internal reports during 1982. The former was noted for analysis to represent use of library documents through reprographic service against each user in the population. For convenience, the number of pages of xerox copies of reading materials from the library in a year was rounded off to nearest hundred in the analysis.

9.6.6 *Correlation of Use of Library Documents through Reprographic Service with User-characteristics*

As could be seen from Table 9.13, the use of library documents through reprographic service correlated highly and positively with status (r_s =

0.93), moderately and positively with professional activities and achievements ($r = 0.53$) and significantly related to qualifications, nature of work and specialisation of the space technologists. As the level of qualifications and management/supervision content of the job increases, space technologists have increasingly taken more xerox copies of reading materials from the library. By and large, the use of documents through reprographic service is almost restricted to high status (Table 9.14), postgraduates and doctorates (Table 9.15) either with management/supervision or planning/system analysis jobs (Table 9.16). Particularly, aeronautical and structural engineers, followed by physicists and electrical engineers (Table 9.17), have made more use of reprographic service than others. The percapita reprographic service availed by doctorates, postgraduates and graduates are in the ratio 4:2:1. Naturally the managers/supervisors and planners/system analysts have made reprographic service four times that of others.

9.7. User-interactions with the Library

9.7.1 *Suggestions for New Documents*

A study of user-participation in collection development (Sridhar, 1983) has revealed that as low as less than 25 per cent of the users participated in the process and such a participation is little over 5 per cent for reports and journals and 20.5 per cent for books. Among those who participated, on an average 3.04 reports, 3.04 books per year and 2.17 journals so far are suggested. Such suggestions from users for new documents accounted for about 35 per cent, 28 per cent and 18 per cent, respectively, of books, reports and journals ordered during a year. The study concluded that the reluctance of users, lack of extensive induction program and opportunity to know about new documents published, the time-bound nature of the projects in which most of the users are involved, and the present relatively slow process of document procurement are some of the reasons attributable for low participation of the users in collection development. Consequently, it is suggested to intensify the current awareness services and dissemination of information about new documents to a larger cross section of the users, systematising literature searching and reference services to get regular feedback to acquisition system, active involvement of project teams in indentifying, in advance, the information requirements of new and ongoing projects and TDPs and speeding up of procurement of suggested documents to create an achievement motivation in the users. The last suggestion which is based on what some users during discussion felt about the present procurement delays was further analysed to explore how far procurement delays are acting as demotivators to user-participation (Sridhar, 1985).

The data on actual procurement of on-demand books, by and large, confirmed the delays felt by the users and out of 818 on-demand books ordered during 1983, only 426 (52%) were supplied by September 1984. The average supplying efficiency of exclusive /specialised Indian dealers, foreign vendors and Indian vendors are 82 per cent, 59 per cent and 28 per cent respectively. It was concluded that newer methods of acquiring on-demand books should be explored and constant reviewing of supply position of vendors be made to revise procurement policies.

9.7.2 *Correlation of Suggestions for New Documents with User-characteristics*

The results of association/correlation tests of suggestions for new documents and procurement, reservation and reference services availed with user-characteristics are shown in Table 9.18. The average number of suggestions for new documents, documents procured, reservations made and reference queries made by the space technologists against their status, qualifications, nature of work and specialisation are presented, respectively, in Tables 9.19, 9.20, 9.21 and 9.22.

As revealed in Table 9.18, the number of new documents suggested by the space technologists correlated perfectly and positively with status ($r_s = 1$), almost moderately with the professional activities and achievements ($r = 0.35$) and related significatly to the qualifications, nature of work and specialisation. Even though the relation with qualifications is positive and linear, (Table 9.20), the doctorates who constitute 4.5 per cent of the population have suggested new documents to the library which equalled the entire suggestions of the rest over a year. The number of suggestions for new documents has linearly increased as the management and supervision content of the job increased (Table 9.21). Further, physicists have dominated the collection development activity of the library by suggesting as many as three times that of an average electronics engineer, four times that of a mechanical engineer, seven times that of an aeronautical/ structural engineer and ten times that of a mathematician and an electrical engineer (Table 9.22).

TABLE 9.18 : Results of Association/Correlation Tests of Suggestions for New Documents and Procurement, Reservation and Reference Services availed with User-characteristics

User-characteristics	No. of suggestions for new documents		No. of documents procured		No. of documents reserved		No. of reference queries made	
	$r_s/x^2/r$	*(df)*	$r_s/x^2/r$	*(df)*	$r_s/x^2/r$	*(df)*	$r_s/x^2/r$	*(df)*
Status (r_s)	1.00*	(6)	0.98*	(6)	0.95*	(6)	0.55	(6)
Qualifications (x^2)	111.78*	(16)	193.33*	(12)	125.28*	(20)	44.07*	(12)
Nature of work (x^2)	49.55*	(24)	39.19*	(18)	59.78*	(30)	17.63	(18)
Specilisation(x^2)	33.55*	(20)	36.31*	(15)	63.87*	(25)	9.75	(15)
Experience *(r)*	0.03	(396)	0.06	(396)	–0.05	(396)	0.00	(396)
Professional Activities and achievements Index *(r)*	0.35*	(178)	0.46*	(178)	0.19*	(179)	–0.03	(179)

Key : *, The association/correlation is statistically significant at $p = 0.05$.

TABLE 9.19 : Status Vs Suggestions for New Documents and Procurement, Reservation and Reference Services Availed

Status	Av. no. of suggestions for new documents	Av. no. of documents procured	Av. no. of reservations made	Av. no. of reference queries made
A, B, C & D	2.59	12.41	6.19	0.22
E	1.76	4.57	4.62	0.45
F	1.73	3.20	3.38	0.69
G	1.51	2.11	3.49	0.39
H	0.81	0.49	2.68	0.30
I	0.66	0.60	3.27	0.44
J	0.16	0.11	0.94	0.19
K, L, M & N	0.02	0.02	0.44	0.02
Total	1.05	2.10	2.81	0.34

Note : The status A to N are in the descending order of the hierarchical rank.

TABLE 9.20 : Qualifications Vs Suggestions for New Documents and Procurement, Reservation and Reference Services Availed

Qualifications	Av. no. of suggestions for new documents	Av. no. of documents procured	Av. no. of reservations made	Av. no. of reference queries made
U Undergraduate	0.04	0.10	0.50	0.04
D Diploma holder	0.32	0.46	1.51	0.36
B Graduate	1.42	1.85	4.06	0.50
M Postgraduate	1.67	3.21	4.19	0.41
P Doctorate	4.40	16.19	7.07	0.72
Total	1.09	2.16	2.94	0,36

TABLE 9.21 : Nature of Work Vs Suggestions for New Documents and Procurement, Reservation and Reference Services Availed

Nature of work	Av. no. of suggestions for new documents	Av. no. of documents procured	Av. no. of reservations made	Av. no. of reference queries made
A Management/ Supervision	1.97	7.00	2.15	0.44
B Planning/System Analysis	1.75	4.71	6.17	0.31
C Design and Development	1.50	2.56	5.46	0.54
P Design, Development, Fabrication and Testing (Both C and D)	1.22	0.91	1.41	0.32
D Fabrication and Testing	0.60	1.30	1.25	0.31
E Operational activity	0.38	0.44	2.51	0.20
P Others	0.53	0.28	3.42	0.67
Total	1.19	2.55	3.45	0.41

TABLE 9.22 : Specialisation Vs Suggestions for New Documents and Procurement, Reservation and Reference Services Availed

Specialisation	Av. no. of suggestions for new documents	Av. no. of documents procured	Av. no. of reservations made	Av. no. of reference queries made
A Physics	3.67	6.03	6.28	0.62
C Mathematics	0.37	3.12	4.41	0.07
E Mechanical Engineering	0.95	0.80	1.18	0.34
F Aeronautical and Structural Engineering	0.50	7.66	2.53	0.53
H Electrical Engineering	0.31	1.43	1.56	0.28
I Electronics	1.12	2.18	4.03	0.35
Total	1.21	2.30	3.25	0.37

9.7.3 *The Procurement Service Availed*

The user-participation in collection development in terms of their indents/suggestions for new documents is just half the story. How many such requirements have been met by the library by procuring and providing them to the user is the other half. The analysis of the data regarding the number of books and reports procured for the users in the population during 1982 runs parallel to the analysis of the suggestions for new documents. The number of books and reports procured during the year against specific request has far exceeded the number of suggestions made during the year, due to accumulated orders of the previous years and due to inclusion of procurement intimations sent to members of the library committee for documents screened and approved by them earlier as part of their function.

9.7.4 *Correlation of Procurement Service Availed with User-Characteristics*

Table 9.18 shows an almost perfect positive correlation of status (r_s = 0.98), moderate positive correlation of professional activities and achievements (r = 0.46) and a significant relation of qualifications, nature of work and specialisation of the space technologists with the number of documents procured and provided to them during one year. In particular, highly placed persons (Table 9.19), doctorates (Table 9.20), managers/supervisors followed by planners/system analysts (Table 9.21) and physicists followed by aeronautical and structural engineers and mathematicians (Table 9.22) have consumed the lion's share of procurement service of the library. An average doctorate has suggested for more than 16 documents a year, which is almost four times the number of documents suggested by an average postgraduate and eight times the number of documents suggested by a graduate (Table 9.20). The number of documents suggested to the library by the space technologists has increased linearly with the increase in management and supervision content of the job (Table 9.21).

9.7.5 *The Pattern of Reservations of Lentout Documents*

User-attitude, intensity of need, awareness and perseverance in borrowing a document are among the factors which have bearing on the document reservation activity in a library. A study (Sridhar, 1983) of the document reservation pattern of the IST based on reservations made by them for lentout books and reports has brought home that about one-third of the users have used this service during a year. The frequency distribution of the number of reservations showed that about 12 per cent of the users have made 42 per cent of the total reservations, one user has made almost one reservation per week and four others have reserved one book per fortnight. The average number of reservations per user during a year (among those who availed this service) was 4.25. Further, maximum reservations were made for books acquired during the current year (*i.e.,* 1982) and most of the reservations were for books acquired during last 3 to 4 years (except for few fundamental classic books in the field). Though slightly more reservations were made on recently published documents, the age of document did not have any relation with the number of reservations made.

Pure science books forming 32 per cent of the collection accounted for 11 per cent of the total reservations and applied science and technology books representing 58 per cent of the collection covered 87 per cent of the total reservations. Reservations for books in other main divisions of UDC were almost negligible. The percentage of reservations for books in a subject was more close to the percentage of users specialising in that subject (except mechanical engineering) than the percentage of stock represented by the subject. The majority of reservations were made on books in high technology and fast growing areas such as electronics, data processing and computer science which together represent the major segment of the existing collection within applied science and technology

It is quite natural to speculate that the attitude of the users about a book and their reservation behaviour are influenced by the background of those who suggested/indented the book to the library and by those (in the organisation) interested in it. A controlled experiment was carried out during October 1985 to see how far the decision of a user to reserve and use a book is influenced by the interest of other colleagues in it. A carefully selected 30 new books were added to the new arrivals display of the library as usual but with a tag on 16 of them with the names (and their scections/ divisions /projects) of the persons who originally indented/ suggested them. The same information about indentors of these books was also provided in the weekly list of additions, copies of which were distributed to all sections, divisions and projects just before the display started. The remaining 14 books fairly matching in subject and nature with the 16 books in the experimental group did not carry any such information about indentors either in the weekly list or on the display. After a week's display, there were 25 reservations for 16 books in the experimental group and 14 reservations for 14 books in control group. The average number of reservations per book during the week for books in the experimental group (where interest of other users/colleagues are publicised) was 1.56 as against 1.00 for books in the controlled group. Thus a 56 per cent extra reservations are attracted by making users aware of the fact that other users/colleagues are interested in them. However, there is not much difference in the attitude of the users about books suggested/indented by communication stars, supervisors and section/division heads compared to books suggested/indented by others.

9.7.6 *Correlation of Document Reservation Activity with User-characteristics*

Table 9.18 also reveals that the number of documents reserved by the space technologists correlated positively and almost perfectly with their status (r_s = 0.95), slightly and positively with professional activities and achievements (r = 0.19) and related linearly and positively with the qualifications. Even the nature of work and specialisation are significantly related to the reservation activity of the users.[11] In particular, doctorates followed by postgraduates and graduates (Table 9.20), planners/system analysts followed by design and development personnel (Table 9.21) and physicists, followed by mathematicians and electronics engineers (Table 9.22) have made more document-reservations than others. Doctorates have reserved books and reports twice frequently than that of postgraduates and graduates.

9.7.7 *Literature Search, Short Range Reference and Reader Assistance Services Availed and Their Correlation with User-characteristics*

The interaction of the space technologists with the library for literature search service is limited to higher status (status A to E only) scientists and engineers. The data is too inadequate to indicate any trend and, hence, could not be analysed further. However, the number of times the short range reference queries made and reader assistance service availed by the space technologists over a year is analysed. Table 9.18 shows that the status of the space technologists is moderately and positively correlated (r_s=0.55) to the number of reference queries made. There is also a significant relation of qualifications of users with the number of reference queries made.

9.7.8 *Other User-interactions With the Library*

The user-behaviour within the library in terms of their interactions with various library service and facilities is an interesting and less explored area in user-research. Such a study has to necessarily exclude the nonusers of the population. A limited sample of the space technologists drawn from the population have been observed for their broad behaviour regarding interactions with the library. The observations relate to the patterns of library visits, library traffic, user movement, use of various services and facilities (within the library) and length of stay of the users in the library.[12]

Based on the data recorded on 24 randomly chosen days during October 1984 and January 1985, it was found that on an average 200 users visit the library (on a working day), of which 55 visited the mezzanine floor of the library where current journals, bound volumes of journals, standards, reprints and product catalogues are housed.

Further, 13 purposeful biased sample of the space technologists (other than those who visited library for a very short time for an issue or return transaction or for indenting or collecting xerox copies) were observed during January 1985 for their movement within the library. In this observation, on an average, a user spent 58 minutes on a visit to the library in 4 to 5 units out of 13 such units within the library. The mapping of users movement indicated that, by and large, the users moved in near critical path avoiding zig zag movements (except when they deliberately wanted to pay repeated visits to a particular unit). Eight out of the 13 only moved on to mazzanine floor and three of them came down to ground floor for use of books and reports and to make enquiries at the circulation counter. The users have often made more than one entry/interaction/visit to the following units: circulation counter, book stack and reading area for books and reports.

Table 9.23 presents the average user visits to ten service areas (units) within the library together with average length of stay. As could be seen from the table, book stack and circulation counter have accounted for more than half of the total visits (56.8%). Display of new arrivals and reading area for books and reports have been moderately visited by the users. A maximum average time of 27.1 minutes per visit has been spent at the display of new arrivals of journals and current journals followed by 20 minutes in the reading area for books and reports and 16 minutes in the bound journals, standards, reprints and product catalogues unit.

About 30 randomly selected Indian space technologists were observed for their length of stay in the library during December 1984 and January 1985. Counting was done from the time the user stepped into the library till the time he stepped out. In order to consider serious inhouse users, all

TABLE 9.23 : User-visits to Service areas within the Library and Length of Stay

Sl. No.	Service Area	Av. no. of user-visits per day	percentage of total visits	Average length of stay (in minutes)
01.	Circulation counter	170.2	27.7	3.1
02.	Reprographics service	17.3	2.8	
03.	Micrographics unit	3.0	0.5	N.A.
04.	Card catalogues	32.0	5.2	8.0
05.	Report stack	6.4	1.0	10.0
06.	Book stack	178.8	29.1	12.0
07.	Reading area (books and reports)	67.0	10.9	20.0
08.	Display of new arrivals (books, reports and standards)	82.5	13.4	5.3
09.	Display of new arrivals (Journals) and current journals.	44.5	7.3	27.1
10.	Bound journals, standards, reprints and product catalogues	12.0	2.0	16.0
	Total	613.7	99.9	—

Key : NA, Not available.

TABLE 9.24 : Number of Users Vs Mean Length of Stay of Users in the Library

Time in minutes (t)	No. of users (N)	N(t) (No. of users staying longer than *t* minutes)
30	9	30
45	8	21
60	6	13
75	2	7
90	2	5
120	2	3
180	1	1
Total	30	30

TABLE 9.25 : Frequency Distribution of Length of Stay of Users

Time in Minutes	No. of users		
	Ground Floor	Mezzanine Floor	Total
<20	4	3	7
20–30	8	0	8
30–40	7	5	12
>40	5	13	18
Total	24	21	45*

Key : *, Total does not tally to 30 due to 15 users using both the ground floor and the mezzanine floor.

those who visited either for reprographic service or exclusively to select and borrow documents were excluded. As such no one in the sample had spent less than 20 minutes.

One hundred and eighty minutes was the maximum length of stay by one user, followed by 120 minutes each by two users. The average length of stay of the users based on mean of the ranges is 58 minutes. About 58 per cent of the respondents spent 5-30 minutes per visit in the library in Raitt's (1984, p.242) study. The number of users and their length of stay in the library based on mean of the ranges are given in Table 9.24. The data in Table 9.24 also indicates that only 7 out of 30 users stayed beyond an hour in the library (in Raitt's study (1984, p. 242), 15 per cent of his respondents spent more than one hour in the library). When this data was plotted on graph, *i.e., t,* the time in minutes plotted against *N (t)* the number of users staying longer than *t* minutes, it resulted in a reverse J shaped curve (Bush, *et al.,* 1956).

One interesting feature is that the average time spent by the users in journals section is much longer than that spent in books and reports sections despite the fact that fewer users have used journals than books. It may be noted (*) that the total number of users in Table 9.25 is shown as 45 (*i.e.* 15 in excess of sample), as they were common users of both ground floor as well as mezzanine floor service areas during a visit. In other words, almost 50 per cent of those who used books also used journals when they visit the library.

NOTES

1. Since data are collected in a series over a period of almost two years, care has been taken to exclude partially represented cases including those who resigned during the survey and those absenting for a long time either on leave or on tour. The data do not represent the complete use or interactions of all the users of the ISAC library as data pertaining to the users not covered in the defined population of the study are excluded.
2. The frequency distribution-pattern of procurement service and suggestions for new documents are almost identical.
3. The multiplying factors or weightages are arrived at based on the relative behaviour of frequency distribution shown in Table 9.1 and the observations such as average number of documents a user has consulted in the library per visit. For example, the frequency distribution of number of pages of xerox copies of reading materials taken is close to that of other types of use of library documents when reduced by a factor of 100 times. Borrowing a document through inter-library loan not only involved extra efforts on the part of the user, but also expected the user to wait for sometime, and the user made a fairly thorough use of such documents compared to documents of the primary library. Users are found to consult on an average 5 documents per visit/ reading seat occupied in the survey. The interactions for literature search are more prolonged as compared to other interactions.
4. Women space technologists did not differ from men in making borrowed use of library documents but differed significantly in inhouse use and interactions with the library (Sridhar, 1987).
5. Those who borrowed more books tend to depend more on formal sources of information ($x^2 = 22.0$, $df = 8$, $p < 0.05$), spend more time on information-gathering activities ($x^2 = 20.8$, $df = 10$, $p < 0.05$), get more bibliographic information from reading materials ($x^2 = 10.8$, $df = 2$, $p < 0.05$) but not necessarily depend more on books for information. This clearly indicates that the degree of dependence on a source of information cannot be equated to numerical figures of use.
6. Unlike books, the number of reports borrowed by an IST is significantly and positively

related ($x^2 = 11.1$, $df = 4$, $p < 0.05$) to his degree of dependence on technical reports for information.

7. Those who borrowed/used more journal issues and even Journal titles tend to spend more time on information-gathering activities, get more bibliographic references from citations in current reading materials but not necessarily depend more on journals for information and not necessarily use more abstracting and indexing journals.
8. The number of books, reports and journals (issues) borrowed by women space technologists did not differ from that of men. However, the average number of journals (titles) used by men and women space technologists respectively are 0.87 and 1.13 (Sridhar, 1987). Surprisingly, those who undertook part time studies do not borrow significantly more books than others ($x^2 = 3.71$, $df = 4$, $p > 0.05$).
9. Like experience, age is also found to be independent of use of books ($r = 0.08$), journals ($r = -0.03$), reports ($r = 0.12$), journal titles ($r = -0.04$) and total borrowed use of documents ($r = -0.09$). Even chi-square test of the use of books, reports, journals and number of journal titles in relation to experience (00–04, 05–12 and 13–30 years) and age (18–34, 35–46, 47–60 years) did not show any significant relation.
10. Interestingly, respondents who underwent part-time studies did not make significantly more inhouse use of the library than others ($x^2 = 2.38$, $df = 6$, $p > 0.05$).
11. Women space technologists have made significantly more reservations than men ($x^2 = 14.7$, $df = 5$, $p < 0.05$) (Sridhar, 1987). Expectantly, those who underwent one or more part- time studies have reserved significantly more documents (average 10.54) than others (average 7.9) ($x^2 = 14.77$, $df = 5$, $p < 0.05$).
12. It may also be noted that, by and large, the users who have made more use of library documents also tend to make more interactions with the library and *vice versa*. Those who made more use of one type of document or more interactions of one type with the library tend to make more use of other types of documents or more interactions of other types. It is the non-users of the library who have remained non-users of almost all the services and facilities of the library tending eventually to become isolated group of nonusers. The library-use index of the IST has correlated moderately and positively ($r = 0.53$, $df = 393$, $p < 0.05$) with their library-interaction index.

10

SUMMARY OF THE FINDINGS, THEIR IMPLICATIONS AND SUGGESTIONS FOR FURTHER RESEARCH

10.1 Realisation of Objective 1 : Information-seeking Behaviour of the Indian Space Technologists

10.1.1 *Motives and Purposes of Seeking Information*

Unlike scientists, the space technologists are not motivated much by recognition, competition, visibility among peers and urge to write and publish (in the decreasing order of priority) while seeking information. On the other hand they are primarily motivated by a need for self-improvement, the desire to be up-to-date in the field of specialisation, maintaining professional competence, self-satisfaction and achieving the desired result in work (in that order).

Keeping up-to-date with latest developments in the field of specialisation/activities and relating present work with the current body of knowledge are the main purposes of seeking information by the space technologists. The space technologists do not seek information much for the purpose of data treatment and processing, crystallising broad and vague assertions, participating in seminars and conferences, writing papers and delivering talks (in that rank order).

10.1.2 *Nature and Type of Information Required*

In consistency with the high ranked motives and purposes of seeking information, the requirements of S&T news and basic S&T information have ranked high for the space technologists. In addition, the space technologists seek more of theoretical background, experimental results, methods, processes and procedures, product, material, equipment and apparatus information and physical, technical and design data (in that rank order) than state-of-the-art, review literature, standard and patent specifications. Information is sought only when it is utmost needed and there is a strong need for internally generated information and product information. Different types of information are used intermittantly in an intermixed way by the IST depending on the nature of work. Change in the nature of work of the space technologists is found to affect both the intensity of need and the type of information required. The problem of 'information overload', enlargement of boundaries of interest and intensification of space activities in the country, impact of electronics boom and resultant need for condensed information-services and need for fast serving information intermediaries or agents are felt by the IST.

10.1.3 *Sources of Bibliographic Information Used*

Library catalogues, colleagues and fellow professionals, direct browsing of library shelves, experts in the field, citations in current reading materials are (in that order) major sources of bibliographic information to the space technologists. The intermittant use of various sources of bibliographic information and interdependence among the sources, indirect cyclic way of generating useful references and more practical way of directly searching the shelves of the library are adopted by the IST for bibliographic references.

10.1.4 *Delegation of Information-gathering Work*

Just 6 per cent of the space technologists have frequently, and another 40 per cent moderately or occasionally, delegated information-gathering work to juniors and subordinates, mainly due to lack of time and/or for involving team members. The tasks delegated are mainly collection of factual data, documents/literature and latest references on specific topics. Those who do not have assistants and those who believe strongly in serendipity value of information-searching process have naturally tend not to delegate information-gathering work. Majority of the space technologists do regular unfocussed browsing of current literature which lead them, at times, to highly useful accidental or chance acquisition of needed information. Respondents have narrated specific incidents of accidental acquisition of inforamtion while skimming through the current journals, unintentionally looking here and there, browsing library shelves and reorganising personal and departmental collection.

10.1.5 *Time Spent on Information-gathering Activities*

The IST have spent, on an average, 9.2 hours per week or one-fourth of their working time for gathering work-related information. Nearly 80 per cent of the space technologists spent 4-15 hours per week on information-related activities. As many as 86 per cent of the IST have spent more than half of such time in seeking/gathering information from formal and documentary sources of information. They could not spend as much time and efforts as they wished for systematic information-gathering. By and large, more time is spent in the initial phases of projects for information-gathering (except where the project is a feasibility study or the project itself is defined and assigned from outside the group) than during later stages. The reverse is true with engineers not involved in conception and definition of projects.

10.1.6 *Dependence on Sources of Information*

The space technologists indicated slightly more dependence on formal and documentary sources of information compared to informal and inter-personal sources. Among the formal sources, they depended heavily on journals followed by books, reports, trade literature and internal reports in that rank order. The IST have qualitatively stressed the importance of trade literature, technical reports, data sheets, conference papers and proceedings, personal and departmental collections and even newspapers. Personal and departmental collections of the space technologists consist mainly of internally generated information, free reading materials, preprints, journals, trade catalogues, standard specifications, reprints, reports, diagrams, handbooks, manuals and are not generally well organised. Among the informal sources, the space technologists have depended heavily on two intra-personal sources, namely, personal experience and results of their own experiments. This is followed by dependence on superiors, peers and colleagues in the organisation. The IST have depended least on fellow professionals outside-the-organisation for work-related information. The nucleus of information-gathering activities of the IST is often oneself. The inter-personal contacts are mainly for opinion, suggestions and references.

Some ways of blending formal and informal sources of information are (1) organisation of internally-generated information (2) inhouse technical journal (3) databank of expertise (4) component co-ordination group (5) inhouse seminars (6) liberal tours (7) presentations from outside speakers and (8) journal clubs.

In order to keep uptodate in the field, over 52 per cent of the respondents depend mainly on journals, another 13 per cent on discussion with colleagues and 9 per cent on trade literature. In addition seminars, newspapers, preprints, lectures and sales representatives are considered quite useful for keeping uptodate. Discussion with similar minds and conscious reading are the two activities which stimulate new ideas to the IST. The tolerable delay in supplying information and documents ranged widely depending on specific instances. Many instances of late detection of information are pointed out by the respondents. Depending on the specific field of work, the technological gap is found to be 5–15 years. The respondents are, by and large, fairly satisfied with the sources of information presently available to them. But they are optimistic on their devoting increasingly more time in future to explore various sources of information.

10.1.7 *Inter-personal Information-sharing*

There exists greater and more free inter-personal communication among peers and colleagues than among subordinates and superiors. There is a general lack of confidence or trust among the respondents about their superiors freely sharing information eventhough many depend on superiors for information and even superiors claim that they collect information for the purpose of sharing it with other members in the team.

10.1.8 *Intra and Inter-organisational Communication*

A very high intra-organisational communication exists among the space technologists as 87 per cent of the five most often-chosen persons for information are from within the organisation and less than one-third of the respondents kept absolutely no inter-personal contacts outside-the-organisation. The most often-contacted persons outside-the-organisation for information are spread over 50 organisations within the country and 22 outside the country. A large majority of communication outside-the-organisation is confined to the persons within the same city and nearly half of such communication is with persons in the academic institutions, where the space technologists have had their education earlier.

10.1.9 *Informal Communication Behaviour*

The communication network based on five most frequently-chosen persons for information revealed that on an average a space technologist is chosen by 3.23 respondents (*i.e.*, average inter-personal information-potential is 3.23). An average 'technocrat' communicated (informally) most often with more than double the number of persons an average non-technocrat communicated. Those with 6-12 years of experience have maximum informal communication ($r = 0.27$). Status or organisational rank is a self-created barrier in inter-personal information-seeking as nearly three-fourth of the contacts are with higher-status persons but status is not a barrier in providing information. Similar barriers do exist in the form of official organisation structure especially divisions, as just one-fourth of the contacts are outside-the-divisions. There exists substantial cross fertilization of ideas as almost half of the inter-personal communication are among persons of different specialisations and fields of activities. Doctorate space technologists are high communicators (informal) eventhough equal number of 'stars' as well as 'isolates' exist among them. Doctorates (30.4 per cent) are twice likely to receive communication from outside-the-organisation than others (15.3 per cent) and 50 per cent more than 'communication stars' (21.6 per cent). Information-potential of an average doctorate space technologist (5.5) is double that of an average person in the population (2.99). One-fourth of the frequent contacts of doctorates for information is with other doctorates.

10.1.10 *Communication Stars and Technological Gatekeepers*

Forty four 'communication stars' (including six 'technological gatekeepers') in the response population have maintained nearly half (47 per cent) of their informal communication among themselves. They have marginally higher communication (22 per cent) outside-the-organisation. The 'stars' are characterised by higher age, experience, status, qualifications, professional activities and achievements, use of library and interactions with the library. They are predominately male technocrats with a confidence of self sufficiency and satisfaction with respect to information-use and transfer and more willing to share information with others. Communication stars do not differ from others in their job satisfaction but they acquire promotion more frequently than others.

10.1.11 *Formal Communication Behaviour*

The formal communication through internal reports has been a more popular method than publishing research papers in journals as space technologists of different status, qualifications and nature of work are involved in preparing internal reports. The space technologists have shown an inclination for oral conference presentation of papers rather than archival publications in journals. Indian journals have served more as publishing outlets for the space technologists than sources of information and sources for citation.

The ranking of various forms of literature in citations of the IST is journals (61 per cent), books (14 per cent), reports (10 per cent), conference papers (10 per cent), theses (2 per cent), preprints (2 per cent) and lecture notes, standards, product catalogues and private communications (1 per cent). Citations to reports and self-citations (16 per cent) are on a higher side compared to results of other studies. Almost 97 per cent of journal articles and 79 per cent of reports cited by the IST are of foreign origin.

The IST have a high coauthorship and collaboration in publishing and research as 81 per cent of the sample papers are multi-authored papers with a 'multiplicity index' of 3.17 and 'productivity ratio' of 0.32. There exists sufficient vertical as well as horizontal coauthorship and collaboration as nearly 80 per cent of coauthorship relations are between persons of different status. Further, 42 per cent and 52 per cent of coauthorship relations are respectively between persons of different divisions/section and different fields of activity.

10.1.12 *Use of the Library and User-interactions with the Library*

The 'primary library' is used by a large majority (88 per cent used library docuements and 60 per cent interacted with library). But it is mainly for keeping up-to-date rather than for increasing their work performance. The space technologists are aware of the consequences of using as well as not using formal sources of information like the library. While the theoretical and basic science oriented persons considered 'papers' as their 'backbones', developmental persons felt that the information from written documents are not often directly applied in the work but are used to have an intuitive guess about future work. After a certain basic level of use, the amount of use of the library is not commensurating with the results of work. The innovative ideas cannot always be put into use due to risk, reliability and cost factors.

The frequency-distribution of use of journals is less skewed than that of books, and books than that of reports. In other words, the interest of the

IST is more widely spread among different journals than in case of books and more widely among books than in case of reports. Unusually, more use of the library is made during reviews for promotion although the space technologists do not attribute seeking information for getting promotion as a major motive or purpose. The borrowed use of books is quite intensive (80 per cent), journals moderate (14 per cent), reports marginal (5 per cent) and standards negligible (0.2 per cent). The inhouse use of journals is much more than borrowed use. Though the use of journals is much less than that of books, the average time spent by the users in journals section of the library is much longer than that spent in books and reports area. Interestingly, there exists a statistically significant moderate positive rank order correlation ($r_s = 0.55$) between inhouse use of library documents and the nearness of the functional division/ project of the user to the library. The average inhouse use by project personnel is maximum and that of the service sector personnel is minimum.

Interestingly enough, the chances of reserving a document by the IST increases by 56 per cent when the interest of colleagues in the new books is publicised. A user who used library documents has greater chances of interacting with the library for other purposes and vice versa. Even a user of a single type of document has a greater chance of using other types of documents and one who interacts with the library for one purpose/service has a greater chance to interact for other purposes/services.

10.2 Realisation of Objective 2 : Characteristics of the Indian Space Technologists

The Indian Space Technologists are working in a young organisation and are themselves quite young (the average age is 31 years). The response population of the study consists of 6.5 per cent females. The average experience of the respondents in ISRO, and before joining ISRO, are respectively, 5.9 years and 2.5 years. A large majority of the respondents brought prior experience from industry (41 per cent), R&D organisations (30 per cent) and academic institutions (18 per cent).

The space technologists are motivated mainly by 'self-improvement' followed by 'work progress' in seeking information. Of the respondents, nearly three-fourths (73 per cent) are fairly or fully satisfied with their present jobs. They have shown a high job mobility within and outside-the-organisation. The performance rating by peers has mostly coincided with the promotion based performance measure. Based on frequency of promotion 158 'high performers' are identified. The frequency distribution of performance based on professional activities and achievements shows that about 13 per cent claim zero value, 37per cent score a low value of 1 to 5, 45 per cent score 6 to 10 and the rest of 5 per cent a high value of 11 or more.

Nearly 55 per cent of the population are scientists and engineers by designation and the rest are non-gazetted technical staff. About 12 per cent of the space technologists are technocrats and a little over 5 per cent work exclusively in project management offices and core teams. Of the respondents, over two-thirds (68 per cent) are involved in design, development, fabrication and testing activities and about one-fifth (20.5) in management, planning and system-analysis work. Large-scale developmental work has necessitated more middle and lower-level and less qualified personnel, particularly in various service sectors. The space technologists show a wide professional diversity in terms of their specialised fields of activities. A majority of the respondents work in the areas of mechanical engineering, communication engineering, computer science, instrumentation and physics. The population consists of just 4.5 per cent doctorates. On the other hand undergraduates and diploma holders constitute 40 per cent of the population. By the criterion of subject of specialisation, there are 44 per cent electronics engineers, 27 per cent mechanical engineers, and 8 per cent physicists. Nearly two-third of the respondents are from applied science, engineering and technology, one-eighth from pure science and a meagre 1.3 per cent from management disciplines. The low percentage of doctorates as well as pure science personnel indicate lack of 'pure research' nature in the work of the space technologists.

By and large, the IST is a heterogeneous group in terms of their characteristics. But the response-population is highly representative of the total population of the study by almost all characteristics. Many user-characteristics of the IST are highly related to each other and status, nature of work, qualifications, specialisation and performance (based on professional activities and achievements) fairly respresent the characteristics of the IST for a correlational study of ISB.

10.3 Realisation of Objective 3: Correlation of Information-Seeking Behaviour of the Indian Space Technologists with their Characteristics

A summary of correlation/association test results of various aspects of ISB with status, qualifications, nature of work, specialisation, experience and

TABLE 10.1 : Summary of Correlation/Association of Information-Seeking Behaviour with six Characteristics of the Indian Space Technologists

Information-Seeking Behaviour	Status	Qualifications	Nature of work	Specialisation	Experience	Professional Activities and Achievements
1	2	3	4	5	6	7
Motives of Seeking Information						
A Own research	I	I	R			I
B Visibility among peers	I	R	R			I
C Edge over competitors		R	R			I
D Recognition		R				
E Promotion		D		R		D
F Project review				R		
G Self-improvement		R				
H For knowledge			R			
I Professional competence					I	I
J Desired result in work	I	I				I
K Publishing papers	I	I	R	R		I
L Continuing education		R		R		I
M Self-satisfaction	I		R			
Purposes of Seeking Information						
A Planning	I	R	R	R		I
B Product/component design and development	I	R	R	R		
C Equipment setup and use	D	R	R	R		D
D Data treatment and processing	I	I	R	R	D	I
E Checking available results/data	I	I	R	R		I
F Checking ones/own results	I	I	R	R		I
G Writing paper and delivering talk	I	I	R	R		I
H Participating in seminars	I	I	I	R		I
I Guiding team	I	R	R		I	I
J Broadening area of attention	I	I	R	R		I
K Keeping abreast with developments	I	I	I	R		I
L Orienting work to existing body of knowledge	I	I	R	R		I

Contd.

TABLE 10.1 (Contd.) :

Information-Seeking Behaviour	Status	Qualifications	Nature of work	Specialisation	Experience	Professional Activities and Achievements
1	**2**	**3**	**4**	**5**	**6**	**7**
M Crystallising broad and vague assertions	I	I	I	R		I
N Evolving innovative ideas/ techniques	I	I	I	R		I
Nature and Type of Information Sought						
A State-of-the-art	I	I	I	R	I	I
B Theoretical background		R			D	
C Experimental results	I	R		R		
D Methods, processes and procedures		R	R			I
E Product, material and equipment information	I	R		R		
F Computer Programs and model building information	I	I	R	R	D	I
G Standard and patent specifications	D	R	R	R	I	D
H Physical, technical and design data	I					
I S & T news				R	I	
J Statistical, economic, business and general information	I	R	R	R		
Sources of Bibliographic Information Used						
A Library catalogues	D	D	R	R		
B Abstracting and indexing journals		R		R		
C Bibliographies and review literature	I	R	R	R		I
D Current awareness bulletins of the library	I	R				
E Citations in current reading materials	I	I	R		I	I
F Announcements from publishers and booksellers				R		
G Browsing library shelves	I		R			
H Browsing in bookshops		R	R			
I Consulting experts		R				D
J Consulting colleagues						

Contd.

TABLE 10.1 (Contd) :

Information-Seeking Behaviour	Status	Qualifications	Nature of work	Specialisation	Experience	Professional Activities and Achievements
1	**2**	**3**	**4**	**5**	**6**	**7**
K Consulting library staff	D	D	R			D
Delegation of Information-Gathering work						
1. Degree of delegation	I	I	I			I
2. Reasons for delegation						
A. Lack of time						
B. Task is tedious and monotonous						
C. Unfamiliarity with the sources	D	D				D
D. Involving team members	I	I	I		I	
3. Nature of information-gathering work delegated						
A. Retrospective and exhaustive search for references					I	I
B. Current search for latest references						I
C. Factual data			R			
D. Documents/literature		R	R			
E. Analysed and digested information						
4. Reasons for non-delegation						
A. No assistants			R			
B. Difficulty in communicating needs						
C. Do not believe in others collecting information						
D. Difficult to train others	I					
E. Do not want to disturb others			R			
F. Searching process is as important as information itself	I	I	R			I
Time Spent on Information-Gathering Activities	I	I		R	D	I
Percentage of time spent on informal and interpersonal sources of information					D	
Dependence on Sources of Information						
1. Formal and documentary sources						

TABLE 10.1 (Contd) :

Information-Seeking Behaviour	Status	Qualifications	Nature of work	Specialisation	Experience	Professional Activities and Achievements
1	**2**	**3**	**4**	**5**	**6**	**7**
A. Books			R		D	
B. Handbooks/reference books		R		R		
C. Conference proceedings and papers	I	I	R	R	I	I
D. Theses and dissertations	I	I	R	R		I
E. Journals		I			I	I
F. Technical reports	I	R				
G. Standard and Patent specifications	D	R	R	R	I	D
H. Official documents and internal reports	I	R	R	R	D	
I. Reprints and preprints	I	I	R	R	I	I
J. Trade literature			R	R		D
K. Personal collection	I	I	R	R		I
2. Informal and interpersonal sources						
M. Personal experience	I	I				
N. Superiors	D	D	R	R	D	D
O. Peers and colleagues in ISRO						I
P. Subordinates and juniors	I			R	I	I
Q. Fellow professionals outside ISRO	I	I	I	R		I
R. Results of own experiments	I	R	R	R		I
S. Professional meeting and seminars	I	R	I		I	I
T. Educational and training courses	D	R			D	
3. Relative dependence on informal and interpersonal sources compared to formal and documentary sources	D	D	D	R		D
The Communication Behaviour						
1. Information-sharing by individuals in ISAC						
A. Subordinates and juniors	I	I	R		I	I
B. Peers and colleagues		R				I
C. Superiors						
D. Others				R		I

Contd.

TABLE 10.1 (Contd) :

Information-Seeking Behaviour	Status	Qualifications	Nature of work	Specialisation	Experience	Professional Activities and Achievements
1	**2**	**3**	**4**	**5**	**6**	**7**
2. Number of regular interpersonal contacts within ISRO	I	I	I			I
3. Number of regular interpersonal contacts outside ISRO	I	I	I		I	I
4. The chances of becoming a 'communication star'	I	I	I	R	I	I
5. Interpersonal information potential	I	I	I		R	I
Use of the Library and User-Interactions With the Library						
1. Library-use Index	I	I	R	R	D	I
A. Borrowed use of books	I	R		R		
B. Borrowed use of reports	I	R	R	R		I
C. Borrowed use of journals	I	R	R	R		
D. Total Use (A+B+C)	I	R	R	R		I
E. Number of journal titles used	I	R	R	R		
F. Inhouse use	I	R	R	R		I
G. Interlibrary loan use	I	I		R		I
H. Use of reprographic service	I	I	I	R		I
2. Library-interaction Index	I	R	R	R		I
A. Suggestions for new documents	I	I	R	R		I
B. Procurement service availed	I	I	R	R		I
C. Reservations made for books and reports	I	I	R	R		I
D. Reference service availed	I	R				
Total number of aspects with which significant relation exist (out of 108 aspects subjected to test)	72	81	67	57	28	66

Key : I-Increases linearly, D-Decreases linearly, R-Related significantly, * - In the increasing order of management and supervision content of the job professional activities and achievements of the IST is given in Table 10.1. Wherever a statistically significant relation exists (R), the same is marked, and, if the relation is linearly related, it is indicated as increasingly/ positively (I) or decreasingly/negatively (D) related, as the case may be. As could be seen from the table, almost all the aspects of ISB are related to one or more characteristics of the IST. Though experience of the IST is related to their ISB, only in 28 of the 108 aspects, other-characteristics relate significantly in 57 to 81 aspects. Thus the null hypothesis that there is no significant relation, other than due to chance, between the ISB and the characteristics of the IST is, by and large, rejected. Hence it can be concluded that the ISB of the IST varies significantly with status, qualifications, nature of work, specialisation and professional activities and achievements.

A mojority of the significant relations of the status of the IST with information-seeking activities are positive. However, seeking information for the purpose of equipment setup and use, need for standard and patent specifications, consulting library card catalogues and library staff for bibliographic references, chances of delegating information-gathering work due to unfamiliarity with the sources, dependence on standard and patent specifications, superiors and educational and training courses as sources of information and relative dependence on informal and inter-personal sources compared to formal and documentary sources of information decrease as the status of the IST increases.

Even in case of qualifications of the IST, most of the correlations/ associations are positive. However, the motive of gaining promotion while seeking information, consulting library card catalogues and library staff for bibliographic references, delegating information-gathering work due to unfamiliarity with the sources, dependence on superiors as sources of information and relative dependence on informal sources compared to formal sources of information decrease as the level of qualifications of the space technologists increases.

The nature of work of the IST arranged in the ascending order of management/supervision content or descending order of operational activity of the job is highly interrelated to their status structure and level of qualifications. As such, the relation of the nature of work to information-seeking activities runs almost parallel to that found with status and qualifications. However, a strict linear positive relation to the ascending order of management/supervision content of the job is found in the case of seeking information for the purposes of participating in seminars, keeping abreast with latest developments, crystallising broad and vague assertions and evolving innovative ideas/techniques, seeking state-of-the-art and review literature, degree of delegation of information-gathering work, involving team members as a reason for delegation, dependence on fellow professionals outside-the-organisation and professional meetings and seminars as sources of information, the number of inter-personal contacts both within and outside- the-organisation, communication stardom, inter-personal information-potential and use of reprographic service of the library. Yet, the relative dependence on informal sources compared to formal sources decreases as management/ supervision content of the job increases. Interestingly, the amount of time spent on information- gathering activities is independent of the nature of work of the IST.

The IST is a heterogeneous group by considering the subject of specialisation of its members. Some of the information seeking activities of specialists within the IST differed from one another. Physicists, mathematicians, mechanical engineers, aeronautical and structural engineers, electrical engineers, electronics engineers have their own profiles of information-behaviour. Physicists, being scientists, are highly active in seeking information and are oriented to literature and formal information-system. Mathematicians are also active seekers of information but not to the extent of the physicists. Among engineers, aeronautical and structural engineers are most active in seeking information and they are close to scientists in their information-behaviour. Out of the rest, electronics engineers are more active seekers and users of information than mechanical engineers and electrical engineers within the IST. As could be seen from Table 10.1, the specialisation of the IST has no bearing on the delegation of information-gathering work. Even the motives of seeking information, use of sources of bibliographic information and communication-behaviour are almost independent of specialisation of the space technologists.

As mentioned earlier, the experience of the IST is found to be correlated with just one-fourth of the information-seeking behaviour items tested for correlation. Interestingly, experience correlated positively with building professional competence as a motive for seeking information, guiding the team as a purpose of seeking information, need for state-of-the-art and review literature, standard and patent specifications, S&T news, dependence on citations in current reading materials, involving team members as a reason for delegation of information-gathering work, dependence on conference proceedings and papers, journals, standard and patent specifications, reprints and preprints, subordinates, professional seminars as sources of information, inter-personal information sharing by subordinates, number of inter-personal contacts outside-the-organisation and the chances of becoming communication star. The aspects negatively related to the length of experience of the IST are : data treatment and processing as a purpose of seeking information, need for theoretical background, computer programs and model building information, time spent on information-gathering activities, percentage of time spent on informal sources of information, dependence on books, internal reports, superiors, educational and training courses as sources of information and library use index. Yet experience is not significantly related to majority of the motives and purposes of seeking information, use of bibliographic sources, delegation of information-gathering work, use of library documents and interactions with the library of the IST.

The professional activities and achievements-index developed in the study appears to be a good measure of the intensity of information-seeking activities of the users. A majority of the information-seeking aspects correlated positively with this index. The aspects of ISB which correlated negatively with the professional activities and achievements index are : getting promotion as a motive for seeking information, equipment set up and use as a purpose of seeking information, the need for standard and patent specifications, consulting experts and library staff for bibliographic references, unfamiliarity with the sources as a reason for delegating information-gathering work, dependence on standard and patent specifications, trade literature, superiors as sources of information and relative dependence on informal sources of information compared to formal sources of information. Yet majority of the nature and type of information sought and bibliographic sources used are independent of professional activities and achievements of the IST.

10.4 Implications of the Results and Findings

It is not feasible to enumerate all the implications of the results and findings of this study. However, an attempt is made here to highlight selected, significant and direct implications of the findings. The multifaceted approach in seeking and collecting information adopted by the heterogeneous group of the IST makes generalisation too difficult. The IST are specialised in different disciplines, yet have a common mission and thereby exhibit the duality of the discipline and mission orientation (Weinberg, 1967) in their information-behaviour and provision of uniform or common service to all is highly unsuitable.

The formal information system should be clear about its objectives (including derived and assumed objectives, if any). Information-services for increasing productivity and enhancing work performance, self progress, departmental promotion, continuing education do differ very much from one another. The excessive dependence on library for departmental promotion and for continuing education has to be taken note of. For example, a text book bank facility either in a centralised or semi-centralised way would not only meet the bulk of such needs of continuing education of the space technologists, but also substantially relieve the burden on central work-related information facility or the 'primary library' of providing information-services to enhance work performance.

Provision of increased opportunity to browse latest literature in both core as well as related/peripheral areas and carrying current awareness services to the workspots and laboratories of otherwise busy IST is necessary. Yet provision for secondary journals, state-of-the-art and review literature can be reduced at the primary library. On the other hand, the library should house more of primary news magazines, trade journals, reference books, data manuals for providing quick and up-to-date information and factual data. More efforts need to be made to provide extensive subject catalogues in the primary library. The resources needed for maintaining author and title catalogues for reports be reallocated for updating the subject catalogues and providing extensive reference service in selection and location of documents by the users.

There is a necessity to provide information- services of the 'primary library' on a regular basis right at the place where the space technologists work. Further, a provision for trained information specialists in an adequate ratio with user-strength be made so that the space technologists can be persuaded to delegate some information-gathering tasks to them whenever they lack sufficient time to undertake such work or whenever their time can be better utilised more productively elsewhere.

A large part of the work-related information needed by the space technologists can be met from a well-organised and comprehensive internally-generated information in the form of technical reports (of various kinds), preprints and reprints of journal articles, conference papers, software documents, design drawings, photographs, slides, transparencies, etc. Since all such information cannot be put at the disposal of all the users, surrogates of such documents should be made available with a note about their location and security clearance, if so needed. The 'reinventing of the wheel' within the organisation can be avoided with such an organisation of internally-generated information.

Due to security reasons or lack of clear policy on security classification many such internally produced documents are not collected, processed and organised at one place within the organisation.

The product catalogues and trade literature, which are considerably sought after by average space technologists are not systematically collected, indexed, organised, stored and updated at one place. Though trade catalogues are normally available free, they are, at times, arbitrarily priced for those who do not buy the concerned product, component or equipment. Enormous trade literature is getting 'buried' in the purchase files denying current information to prospective buyers, and servicing and maintenance personnel. Such literature should be taken care of as multiple requests for free copies of manuals and other related trade literature from the same organisation are normally discouraged by the manufacturers and even when they wish to supply either free or on nominal payment, copies may not be available for equipments and components sold earlier. A fairly centralised data bank of trade literature which acquires, indexes, organises and systematically weeds obsolete product literature is necessary.

There is a need for intermediary information-specialists who can act as data and information-gathering and analysing agents and support the local information-analysis centre. These intermediaries should work closely with the senior space technologists, projects and the primary library. They should screen and sift the reading material to be browsed by the senior space technologists and provide highly reliable and accurate data/information to those who need it.

A process of generating confidence among subordinates and juniors for seeking information from superiors and seniors is necessary to overcome the 'psychological cost' or 'status inhibition' exhibited in the transfer of information. In a highly intra-organisation-oriented communication of the space technologists, there is a need for cutting recirculation of ideas or information among members of tight and closed loop informal network of 'old boys'. The marginal inter-organisational communication due to communication stars and technological gatekeepers and tours and visits abroad (of mostly of higher-status persons) may have to be increased by encouraging more contacts outside-the-organisation especially at middle and lower-levels. A more liberal organisation structure which places communication stars accessible to more number of colleagues and makes many of them 'linking pins' in the organisation structure might optimise use of such high- information potentials. In addition, increasing the ratio of technocrats to nontechnocrats, periodic reassignment of jobs, more encouragement of inter-section, inter-division and inter-project communication bringing 'isolates' into the communication network is needed.

Encouraging and giving due credit for publishing activity is more likely to result in an increased use of literature more frequently and systematically and intensify engineers' information-behaviour.[1] The publishing activity also encourages more coauthorship and collaboration and inturn informal communication. Further, a proper security classification scheme could be evolved to give publication status to all unclassified and declassifed internal reports of the organisation. This not only aids internal generation of information but also motivates authors to be more literature and documentation oriented in production of internal reports. Moreover, the cost of production and distribution of such reports can be recovered by charging those buying such reports and/or by exchanging such reports with reports of other similar organisations.

For the purpose of blending formal and informal sources of information the following options can be considered :

(i) developing 'components databank' and 'components co-ordination group';

(ii) disseminating various technical information and activities through the inhouse technical journal;

(iii) developing a panel of experts on whom technologists in the organisation can look upon for getting authentic information quickly on a given specific area without much 'psychological cost';

(iv) conducting regular inhouse technical seminars, and circulation of brief summaries, detailed summaries and the complete documents relating to such seminars, reviews, etc., to appropriate persons; and

(v) creating more opportunities for the space technologists to meet fellow professionals outside the organisations by allowing liberal tours and/or by inviting outside professionals for presentation of papers/reports in inhouse seminars.

The formal information facility *i.e.*, the 'primary library', needs to be strengthened in respect of nature and type of information needed by the IST. The overall organisation planning in general and planning of various projects and activities (including TDPs), in particular, have to make provision for information-support needed. The members of the projects and groups should be made to develop the habit of participating in the collection-development process of their primary library. It is necessary to see that each section, division and project formulates a systematic way of circulating current awareness services and other surrogates, announcements about documents and summaries among members of their sections, divisions and projects.

The primary library, on its part, should evolve a more liberal policy of promoting the use of journals, reports, standards and trade catalogues among the users concerned without expecting the space technologists to visit the library to consult a document or to reserve a document or to borrow a document for a limited period and, at times, pay a heavy fine as penalty for having not returned in time. This may necessitate a need to be more flexible in the application of library rules and regulations excepting a few areas like textbooks and popular books. For example, nearly three-fourths of the sample reports of the primary library which are not borrowed could be sent to the work tables of the potential users concerned (atleast copies of the title pages, abstracts and document control data sheets) for perusal rather than passively storing them on library shelves. Above all, the primary library has to find the ways and means of inducing and inducting the non-users. There is a need to give orientation programmes to the space technologists (especially to new entrants and less frequent users) in the use of the primary library. The experimental services like the 'pathfinder' and 'Aid to space science and satellite technology orientation' have been well received and efforts need to be continued on similar lines. The space technologists also need an informal training in organising their personal information system consisting of references, notes, abstracts and documents in various forms such as registers, notebooks, files, loose sheets, cards and desk diaries.

It is necessary that the organisation and the primary library take cognisance of various information-behaviour aspects and activities which are shown to be functions of one or more variables (Table 10.1) and accordingly plan, design and operate the information-system. These relations of the user-characteristics to information-behaviour may be kept in view in providing information-services to various groups within the IST.

The professional activities and achievements index developed in this study has fairly represented the intensity of information-seeking, information-orientation, the information-potential and the library/literature-orientation of the space technologists which could be made use of in repatterning the information-services.

10.5 Suggestions for Further Research

The present study is confined to pursuing and achieving the three objectives and testing the hypothesis stated earlier. In this pursuit many related issues/problems were noticed. Though they were delimited as outside the scope of the study, many of them are worth considering them on their own as areas for further research. Here, then, are presented such areas as well as other areas for further research including those of ISB of the IST.

10.5.1 *User-Research in General*

There is a strong need for a meta-analysis[2] (i.e., the analysis of several analyses) of many past user-studies over the last four decades in science and technology areas so that a thorough sifting and stock taking as well as consolidation of the findings of user-research takes place. It is worthwhile to arrive at a generalised user-behaviour research model using the pattern of the present and other past studies as bases so that each one of the major organisations, disciplines and missions within science, engineering and technology in the country could be studied preferably by the information-specialists 'living with the tribe'.[3] Further comparisions and meta-analysis would become meaningful and dependable only if such a research model is adopted for series of studies covering different user populations. Such a model could also be used for periodic information-behaviour studies of the same set of users.

There is also a need for a national level multidisciplinary research organisation in India on par with BASR (ISR), CRUSK, CRUS, etc., of the western countries to take care of continuous research on communication and information-behaviour of the Indian scientists, engineers, technicians and technologists.

The basic aspect of user-research, namely, 'why does the information-seeker behave as he does?', 'Why does ISB vary with user-characteristics?', 'Is ISB contingent to the work atmosphere or dependent on attributes of individuals?' are some of the fundamental areas which demand in-depth studies.

More surveys are necessary not only to confirm general patterns of information-behaviour but also to identify departures from the norm (Hanson, 1974). Some of the areas needing urgent attention are assessing the impact of different kinds and levels of presentation of reading materials, ways of measuring accessibility to information and ascertaining the effects of such accessibility, impact of information on its receipient and on the progress and outcome of his work. In addition, extensive local, organisation, discipline and mission oriented user-research within the country is needed.

10.5.2 *Other Aspects of Information Behaviour*

An intensive multidisciplinary research of motives of seeking information, a study of how unmet needs of users get dissipated, ways of measuring 'information overload' and impact of 'information overload' on use of information are some of the other areas for further research. A thorough review of the need for catalogues and indexes in libraries and their functions in serving engineers and technologists, how they can be made more attractive and useful to users in providing biblographic information, how the library can aid specialists and experts in providing bibliographic information and how subject specialists be developed as information intermediaries for providing information-services so that scientists, engineers and technologists can confidently delegate their information-gathering work to them are the other wider avenues for further research.

More research is also needed about the information-processing behaviour of scientists, engineers, technicians and technologists, i.e., the way they use documents and information as well as relating use of information to the purposes of seeking information and the requirements of information. How the use of information varied at different stages of a work or project, what sources of information under what setting (bringing text and user together) generates new/novel ideas, increases 'accidental acquisition' of information and reduces chances of 'late find' of information need to be explored.

More research by librarians about informal and inter-personal sources of information as well as informal communication of scientists, engineers, technicians and technologists is needed. In particular, the ways and means of increasing effectiveness and efficiency of informal communication have to be explored. A semi-controlled experiment to test the hypothesis that 'high performance causes high communication and not vice versa' would be quite interesting. Similarly, a study to distinguish the effects of formal organisation structure and informal social relationships on communication of scientists, engineers, technicians and technologists is highly desirable. One specific hypothesis could be that 'the effect of informal social relationship is much stronger than the formal organisation structure on communication'. Another potential hypothesis to be tested could be that 'the informal communication among peers and colleagues is much more effective and cordial than between a superior and a subordinate'.

Further, can communication stars and technological gatekeepers be formally appointed? and, if so, how do they differ from self-chosen communication stars and technological gatekeepers in their effectiveness? Such issues require systematic studies. To put it in the form of a hypothesis 'the communication stardom and technological gatekeepership are intrinsic to individuals and not contingent to a particular organisation'.

Another less explored area is that of informal communication among coauthors. There is a need to compare the characteristics of coauthors and also to determine the nature and amount of work shared by coauthors in producing a paper or a document. 'Does co-authorship-relations result in lasting and highly effective informal communication among coauthors?' is quite speculative in nature. A qualitative study of citations which could throw light on how use of information is related to its citation is necessary. Why one cites a reference and whether all the references cited have been read and used by the author are to be answered. Further, research regarding various interactions of scientists, engineers, technicians and technologists with formal sources like a library is needed to assess their real time responses for arriving at their information-behaviour as well as their attitude towards information.

10.5.3 Further Research on Information-Behaviour of the Indian Space Technologists

The present study may be extended to cover other scientists, engineers, technicians and technologists, in general, and space technologists working in sister organisations in particular, in India, so that results could be compared. There is a vast scope for further research in terms of different types of users, different aspects of user-behaviour and attitudes and comparison of user-behaviour and attitudes of different types of users. In addition, the basic research aspect of ISB of the space technologists could also be explored. The information requirements of about 100 industries which worked with ISRO and supplied materials, equipment and services and another 25 industries which used ISRO-developed technology for non-space uses as well as the organisations which undertook sponsored projects and contracts of ISRO are worth probing. Many other user-characteristics/attributes including personality traits and psychological dimensions of information-seeking and using need to be identified and their relation with information-behaviour of the IST tested.

Further analysis of the database created in the present study could also be taken up especially by cross tabulating the data. For example, the nature and type of information sought against the sources of information preferred and ranking of various factors within ISB aspects (like motives and purposes of seeking information, nature and type of information required, dependence on sources of information, etc.) can be compared for different user-characteristics. There is ample scope for finding inter-correlation among various factors within each aspect of information-behaviour as well as inter-correlating various aspects of information-behaviour. Thus, the findings of the present study as well as its database could serve as a starting point for many independent short and long range research projects which may in turn throw better light on the ISB of the IST.

10.5.4 *Use of Refined Methodologies and Data-Analysing Techniques*

From the point of view of methodology and data-analysing techniques, there is a need for the use of semi-controlled experiments, multiple correlation, partial correlation, factor and cluster analysis, analysis of variance, regression-analysis and other techniques. For example, the quantitative data such as age, experience, professional activities and achievements, library-use and library-interactions of the space technologists can be subjected to linear regression-analysis.

NOTES

1. Raitt (1984, p. 275) also make similar observations.
2. Meta-analysis is a methodology quantitatively cumulating and integrating findings across studies (Gooding and Wagner, 1985, p. 465).
3. This should have been the foremost task of the apex bodies like NISSAT in the country.

Appendix **1**

FORMAT OF WEEKLY SUMMARY SHEET FOR INFORMATION RELATED ACTIVITIES ISAC : LIBRARY

No. ISAC/LIB/ISB/82 **December 28, 1982**

NOTE

Sub : Request for data about Information activities

A study of information-seeking behaviour of space technologists has been undertaken with the objective of knowing the users and their information related activities in detail. In this connection, I solicit your valuable help and cooperation in understanding how a typical scientist/engineer/ technical person of ISAC interacts with scientific and technical communication, involves himself in information transfer process and the nature of his behaviour in searching, identifying, acquiring, availing and transferring desired work-related information. Information for this purpose includes oral, written or documentary, statistical, pictorial, descriptive, bibliographic and other forms.

In persuance of this study, collecting (from selected scientists, engineers and other technical staff) on a weekly basis, a summary of information activities covering all significant events related to above aspects for about six weeks would be of immense value. Hence, I request you to provide frank and objective data in the enclosed format. I assure you that the data so provided will be kept strictly confidential.

You may return the filled sheets either to the undersigned or drop in the library suggestion box either on weekly basis or at the end of six weeks.

(M.S. SRIDHAR)
Librarian

To. ______________________________

____

____

SOME SUGGESTIONS FOR WRITING SUMMARY SHEET

You have all the freedom to record in your own way the details of information-related activities in the summary sheet. However, the following suggestions may kindly be considered to facilitate easy processing of the data:

1. Record in detail specific instances of information collection as well as transferring activities restricting the same to work related information.
2. Your general expeiences about availing various information sources may be added. Information sources may be formal/ documentary like official documents, departmental and personal collection, trade literature, reprints and preprints from fellow professionals, books, periodicals, reports and other documents normally available at libraries, etc. Informal/interpersonal sources include personal experience, superiors, peers, subordinates, analysis and results of experiments, professional meetings, seminars, symposia, lectures, educational and training courses, etc.
3. Record how a particular source of information is used. Indicate your degree of dependence and success on different sources, if used.
4. How a particular need for information is met or could not be met may be given as a case study.
5. Record approximate time (in hours) you spend in information-related activities during a week.
6. A hypothetically worked summary sheet is enclosed as an indicative example.

SUMMARY SHEET FOR RECORDING INFORMATION SEEKING ACTIVITIES

(A hypothetical example)

Staff No.

Dec. 4-10, 1982

WEEK 1,

1. I wanted more information on '____________________' especially on application aspect of it and found difficult to get it.
2. My need for information/literature on '________________' could not be satisfied over last few weeks. I gave up further efforts.
3. So and so has provided very useful and ready made data about '______________' for my work which otherwise would have taken enourmous effort and time to collect the same.
4. Attended a lecture on '__________________________' by so and so. It was informative and useful to my work. Incidentally it triggered my interest in '______________' aspect of my work. I had an extensive informal discussion with colleagues in the section on the above topic.
5. While browsing latest literature. I accidentally discovered an interesting piece of information on '__________________' which changed my line of thought.
6. I spent more time in reading than normally I do. in a week. in order to prepare for a lecture on '______________________'
7. I spent about 20 hours during the week on information-related activities, which is abnormally high and the same is attributable to preparation for a lecture. Out of 20 hours, 10 hours have been spent with informal/interpersonal sources (Almost equal time is spent on formal and informal sources)
8. I received one reprint and five trade catalogues during the week and found all of them very useful for my work at ISAC.
9. This week I depended heavily on internal report for my work.
10. I rejoiced when I discovered that the manual borrowed from '____________________________' section gave all the details

for my work on ' ______________ '. This work was held up for long time for want of details on '_'

11. I could not find enough time for information activities as I was more involved in routine work such as ' ________________ '

12. I made extensive preparation to face DPC review and spent more than 30 hours during the week in reading literature. Visited ' ______________________________ ' library.

13. I attended a seminar on ' ______________________________ ' and found the same not much useful.

SUMMARY SHEET FOR RECORDING INFORMATION SEEKING ACTIVITIES

Staff No.

Jan 2-8, 1983

WEEK 1,

Appendix 2

QUESTIONNAIRE USED FOR DATA COLLECTION
ISAC : LIBRARY

No. ISAC/LIB/ISB/83 **June 15, 1983**

NOTE

Sub : Request for data regarding information related activities

1. I have undertaken, with the permission of authorities, 'A study of Information-seeking Behaviour of Space Technologists'. Without confining to the use of libraries, the study aims at understanding how space technologists gather information and participate in scientific and technical communication.
2. You are chosen as one of the representative persons to collect necessary data for the study. Hence I request you to provide through the enclosed questionnaire, valuable data on how you search, identify, acquire, avail, transfer and use work-related information of present as well as future requirements. Results of such a study will be of immense value in planning appropriate information system and services.
3. This is not a fault finding study. Certain personal data such as nature of work, specialisation, experience are requested in the questionnaire only for a correlational analysis of information-seeking behaviour against such data. The name of staff number of respondent is optional and only to facilitate further clarifications and discussions if needed. I assure you that the opinion and data collected from you will be kept strictly confidential and no name will be cited for whatsoever. This being a behaviour oriented study, result will be as objective as the data is. Hence your frank and factual reply will be highly appreciated.
4. For further clarification or assistance kindly dial 202. If you wish to have private, face to face discussion on any aspect of the questionnaire, I shall be too happy to call upon you on knowing your convenience. Kindly return duly filled questionnaire to me

or drop the same in the library suggestion box. Once again, I solicit your kind cooperation in the matter.

(M.S. SRIDHAR)
Librarian

Encl. a.a.
To. ____ ____________________
____
____________________________ __

GUIDELINES FOR FILLING THE QUESTIONNAIRE

1. A careful thought to each question before answering is requested.
2. Information is that which adds to knowledge and/or helps one in decision making and problem solving situations. Information could be written or oral.
3. Sources of Information include formal and documentary sources such as books, periodicals, reports, etc. as well as informal and interpersonal sources such as one's own experience, informal discussion with colleagues, results of experiments, meetings, seminars etc. (see Q. No. 3).
4. Information-seeking behaviour means attitude, ways and means of searching, identifying, acquiring and using desired work-related information of present or future requirements.
5. As far as possible your replies to the questions need to be based on your experiences in information-seeking during last one year.
6. Most of the questions are closed-ended requiring you either to tick (✓) mark or place appropriate code number given along with the question in the left hand side box.
7. There are few questions for which statisitcal data have to be given. You may provide average and/or approximate figures for such questions in the left hand side boxes provided.
   e.g.,

8	E. Number of papers published so far.

8. In case of open-ended questions, let your opinion and comments be brief and unambiguous.

QUESTIONNAIRE FOR A STUDY OF INFORMATION-SEEKING BEHAVIOUR OF SPACE TECHNOLOGISTS

To be filled and returned to :

M.S. Sridhar
Librarian
ISRO Satellite Centre
Bangalore-560 058

1. Please check, on a five-point scale, using the codes given below, the nature and type of information required by you.

CODES

0	Not required
1	Rarely required
2	Occasionally required
3	Frequently required
4	Highly sought after

☐ A. State of the art and review literature

☐ B. Theoretical background/Basic scientific and technical information.

☐ C. Experimental designs, results and applications.

☐ D. Methods, processes and procedures

☐ E. Product, material, equipment and apparatus information

☐ F. Computer programs and model building information

☐ G. Standard and patent specifications and codes of practice

☐ H. Physical, technical and design data

☐ I. Scientific and technical news

☐ J. Statistical, economic, business and general information

☐ K. Other type of information, if any. Please specify below:

2. Please check, on a five-point scale, using the codes given below, the purposes for which you collect and use information

CODES

0 Not used
1 Rarely used
2 Occasionally used
3 Frequently used
4 Highly used

[ ] A. Program/activity planning

[ ] B. Designing and development of products and components

[ ] C. Setting up and use of equipment

[ ] D. Data treatment and processing

[ ] E. Checking authenticity of available results/data/information

[ ] F. Checking and Evaluating ones/own results

[ ] G. Writing papers and delivering lectures

[ ] H. Participating in seminars and conferences

[ ] I. Guiding team and sharing information with members of the team

[ ] J. Broadening area of attention and reviewing work done in related areas

[ ] K. Keeping abreast with the latest developments in the field

[ ] L. Orienting ones/own work with the existing body of knowledge

[ ] M. Crystallising broad and vague assertions

[ ] N. Evolving innovative ideas/techniques

[ ] O. Other purposes, if any. Please specify below:

3. Please check, on a five-point scale, using codes indicated, the extent to which you depend for information on the sources given below:

CODES

0 No dependence
1 Rare dependence
2 Occasional dependence
3 Frequent dependence
4 High dependence

3.1 Formal and documentary sources:

- [ ] A. Books (other than handbooks/reference books)
- [ ] B. Handbook/reference books
- [ ] C. Conference proceedings and papers
- [ ] D. Theses and dissertations
- [ ] E. Periodicals/journals
- [ ] F. Technical/R&D reports
- [ ] G. Standard and patent specifications
- [ ] H. Official documents including internal reports of ISRO and departmental collection
- [ ] I. Reprints and preprints from fellow professionals
- [ ] J. Trade literature/product catalogues from suppliers and manufacturers
- [ ] K. Personal collection
- [ ] L. Other formal sources, if any. Please specify below :

3.2 Informal and interpersonal sources :

- [ ] A. Personal experience
- [ ] B. Superiors
- [ ] C. Peers and colleagues in ISRO
- [ ] D. Subordinates and juniors
- [ ] E. Fellow professionals outside ISRO
- [ ] F. Results of one's/own experiments

☐ G. Professional meetings, seminars, symposia and lectures

☐ H. Educational and training courses

☐ I. Other informal sources, if any. Please specify below :

4. How much do you depend on informal and interpersonal sources compared to formal and documentary sources (see 3.1 & 3.2 above) ? Tick (✓) mark any one.

☐ A. Almost on informal sources only

☐ B. More on informal sources than formal sources

☐ C. Equally on both informal and formal sources

☐ D. More on formal sources than informal sources

☐ E. Almost on formal sources only

5.1 How much time do you normally spend in gathering work-related information, literature and references both from formal as well as informal sources? (Include time spent in literature search, retrieval of references, locating documents, purposive interpersonal communication, etc.). Please tick (✓) mark any one.

☐ A. More than 20 hours per week

☐ B. 16 to 20 hours per week

☐ C. 11 to 15 hours per week

☐ D. 7 to 10 hours per week

☐ E. 4 to 6 hours per week

☐ F. Less than 4 hours per week

5.2 What percent of this time is spent on informal and interpersonal sources

6. How do you get relevant references (Not necessarily information and documents) of your requirements ? Please tick (✓) mark all those applicable.

☐ A. Library catalogues

☐ B. Abstracting and indexing periodicals/journals

☐ C. References from bibliographies and review literature

☐ D. Current awareness bulletins (*e.g.* list of additions) of library

☐ E. Citations in current reading materials such as books, reports and periodical/journal articles.

☐ F. Announcements from publishers and booksellers

☐ G. Browsing in library shelves

☐ H. Browsing in bookshops

☐ I. Consulting experts in the field

☐ J. Consulting colleagues and fellow professionals

☐ K. Consulting library staff

☐ L. Other means, if any. Please specify below :

7.1 Do you delegate the work of information collection to others ? Please tick (✓) mark any one

☐ A. No Delegation

☐ B. Delegate Occasionally

☐ C. Delegate Moderately

☐ D. Delegate Frequently

7.2 If you delegate, check reasons for such delegation by tick (✓) marking all those applicable.

☐ A. Lack of time

☐ B. The task is tedious and monotonous

☐ C. Unfamiliarity with the sources

☐ D. Making team or group members involved

☐ E. Other reasons, if any. Please specify below :

7.3 If you delegate, check the nature of work delegated by tick (✓) marking all those applicable.

☐ A. Retrospective and exhaustive search for references

☐ B. Current search for latest references

☐ C. Factual data such as design data and physical data

☐ D. Documents/Literature

☐ E. Analysed and digested literature/information

☐ F. Others, if any. Please specify below :

7.4 If and when you do not delegate, check reasons for nondelegation by tick (✓) marking all those applicable.

☐ A. Do not have assistants to delegate

☐ B. Find it difficult to communicate my information needs to others

☐ C. Do not believe in others collecting information for me

☐ D. Find it difficult to train others in my ways of collecting infrormation

☐ E. Do not want to disturb others

☐ F. I believe the searching process is as important as information itself

☐ G. Other reasons, if any. Please specify below :

8.1 Generally, with how many persons you have regular contact for information, (*i.e.*, approximate size of informal network accessible to you) ?

☐ A. Within ISRO

☐ B. Outside ISRO

8.2 Please name five most-often contacted persons

Name	Organisation/ Division/Project

A.

B.

C.

D.

E.

9. Mark with the below mentioned codes, extent to which others in your centre/project share information with you.

CODES

0 No information is shared

1 Not much information is shared

2 Certain type of information is shared

3 Most of the information is freely shared

☐ A. Subordinates and juniors

☐ B. Peers and colleagues

☐ C. Superiors

☐ D. Others

10. Comment here briefly on your information-gathering habits and provide any other relevant information you would like to. (Use separate sheet if required)

11. What motivates you to seek and collect information ? Please check on a five-point scale using below mentioned codes.

CODES

0	Non-motivator
1	Weakest motivator
2	Average motivator
3	Fairly strong motivator
4	Strongest motivator

☐ A. To pursue own research related to work/in the field

☐ B. To have visibility among peers and colleagues

☐ C. To have an edge over other competitors

☐ D. For recognition

☐ E. To prepare for departmental review for promotion (DPC)

☐ F. To prepare for project review

☐ G. For self-improvement

☐ H. To acquire and update knowledge in the field

☐ I. To maintain professional competence

☐ J. To achieve desired result in work

☐ K. To write and publish

☐ L. To pursue continuing education

☐ M. For pleasure of doing good work, self-fulfilment and self-satisfaction

☐ N. Other motivational factors, if any. Please specify below:

12. Please mention the extent of your professional activities and achievements.

☐ A. Number of lectures delivered in last one year.

☐ B. Number of seminars symposia and conferences attended in last one year (other than routine internal reviews)

☐ C. Number of times other centres and organisations were officially visited in last one year (*i.e.*, No. of tours)

☐ D. Number of internal documents/reports prepared in last one year.

☐ E. Number of papers published so far

☐ F. Number of patents obtained so far

☐ G. Number of awards and citations (relating to achievements in the profession) received so far

☐ H. Number of professional associations, societies, etc. in which you are a member

☐ I. Number of board of studies or other similar bodies of academic institutions in which you are a member

☐ J. Number of committees (both within and outside ISRO) in which you are a member or chairman

☐ K. Number of editorial committees or boards of technical or professional periodicals/journals in which you are a member

☐ L. Number of professional periodicals/journals to which you personally subscribe (excluding free journals)

☐ M. Part-time studies (for certificate, diploma or degree) undertaken in last two years.

13.1 Tick (✓) mark any one to indicate your present nature of work based on maximum time consumed in your day-to-day work.

☐ A. Management/Supervision

☐ B. Planning/System Analysis

☐ C. Design and Development

☐ D. Fabrication and Testing

☐ E. Operational Activity (involving mostly repetative tasks with no supervision and planning and least of design, development and fabrication)

☐ F. Others if any. Please specify below :

13.2 Is there any substantial change in the nature of your work in recent years (within ISRO). Please tick (✓) mark any one.

☐ A. Yes

☐ B. No.

13.3 What is your present major specialised field of activity?

14.1 What is/are your educational qualification/s ?

14.2 Tick (✓) mark any one to indicate your broad area of specialisation in highest qualification.

☐ A. Physics

☐ B. Chemistry

☐ C. Mathematics

☐ D. Civil Engineering

☐ E. Mechanical Engineering

☐ F. Aeronautical and Structural Engineering

☐ G. Thermal Engineering

☐ H. Electrical Engineering

☐ I. Electronics

☐ J. Computer Science

☐ K. Chemical Engineering/Technology

☐ L. Material Science

☐ M. Management

☐ N. Others, if any. Please specify below :

15.1 Indicate your work experience (excluding apprenticeship) in years.

☐ Yrs. A. In ISRO

☐ Yrs. B. Prior to joining ISRO

15.2 Please tick (✓) mark all those applicable to indicate nature of organisation where worked prior to joining ISRO.

☐ A. Academic/Educational

☐ B. R & D

☐ C. Industry

☐ D. Others. if any. Please specify below :

16. Please tick (✓) any one to indicate the degree of overall satisfaction in your present job.

☐ A. Fully satisfied

☐ B. Fairly satisfied

☐ C. Partially satisfied

☐ D. Not much satisfied

☐ E. Not satisfied

17. Your Division/Project

18. Your Designation/grade

19. Your age (in years)

20 Number of promotions secured in ISRO

Note : If you have no objection, please note your name or staff number to facilitate further clarification and/or discussion when required.

Appendix **3**

INTERVIEW SCHEDULE

INTERVIEW SCHEDULE FOR A STUDY OF INFORMATION SEEKING BEHAVIOUR OF SPACE TECHNOLOGISTS

Interviewer	**Respondent**
M. S. Sridhar	Name :
Librarian	Staff No. :
ISRO Satellite Centre	Divn/Project :
Bangalore - 560 058	Date :

1. To know the sources on which the space technologists depend for latest information :
   How do you keep uptodate with latest developments in your field of specialisation/area of activity ?
2. To know the extent of utility of information collected by space technologists :
   The ultimate aim of collecting work-related information is to achieve effectiveness and efficiency in the results of your work. To what extent the information you have been collecting has helped you in your work ?
3. To know how formal and informal sources of information can be made to supplement each other :
   For your information needs, you make use of many formal and documentary sources such as books, periodicals/journals, reports conference papers and proceedings, internal reports, standards, etc. as well as informal and interpersonal sources such as one's/own experience, discussions with colleagues and fellow professionals, results of experiments, meetings, seminars, etc. If an individual is to gain maximum information with least possible efforts, it is advisable that formal and informal sources have to effectively supplement each other. Keeping this in view, (1) to what extent the formal sources supplement information gathered from informal sources and *vice versa* ? (2) could you suggest how best the formal sources can be made to supplement informal sources and *vice versa* in your case ?

4. To know the nature of information-sharing and strength of informal sources :
Usually considerable amount of information is exchanged in interpersonal discussion among colleagues and fellow professionals. Is this true in your case ? If so, could you let me know (1) what type of information is shared, and how it is shared, (2) How important are these informal sources to you and what is the reliance that you place on such informal sources ?

5. To know variations in time spent on information-gathering activity :
Perhaps, you spend considerable time on gathering information required by you. How does such time allocation (or consumption) for information-gathering activity vary with different stages (conceptual, proposal, initial, middle, final and post completion stages) of your work/project/program?

6. To know how change in nature of work affects information-seeking behaviour : (only for those who have answered Q.No. 13.2 of the questionnaire in affirmative).
During your stay at ISRO, has your nature of work* changed at any time ?

---

1. Management, supervision, etc.
2. Planning, system analysis, etc.
3. Design and development
4. Fabrication and testing
5. Operational activity.

If yes, how this change has affected (a) your information requirements ? and (b) your information-seeking/collecting activities ?

7. To know whether activities such as reading, browsing, information-collection, discussion, thinking etc., play any role in generating creative thoughts or ideas.
Normally, everyone has his/her own 'best time' for getting better 'brainwaves' and doing creative things. In the light of this background, which of your activities provide you better stimuli/impetuses for new ideas or novel thoughts ?

8. To know how space technologists apart from being users of information play the role of information producers and information transferring agents :
Every user of information is also a producer and transferer of

information. How do you rate yourself as a producer and a transferer of information compared to user role ?

9. To know whether document utility is simultaneous or sequential: You read, discuss and use many written documents for extracting information. Do you normally use such documents consecutively or conjunctively and why so ?

10. To know the criticality and tolerance of information need : When you demand for a particular document or a specific information from a formal source such as library, there is a likelihood of certain delay in supply of the document or information. What, according to you, is the tolerable delay in supply of document and specific information needed by you ? In this connection, could you cite and example of 'late supply' or 'late detection' of a document/piece of specific information which did not serve the purpose ?

11. To know technological gap and 'how old' documents are acceptable by users : India being a developing country lags behind developed countries in many technologies. What is the extent of technological gap if any, in years, in your field or area of activity ? Is the gap being narrowed down or widened over years ? How much old (by year of publication) documents and how much old (by year of production) information are acceptable to you ?

12. To know about accidental acquisition of information and exploring means such as 'browsing activity' to increase accidental acquisition : You may come across relevant and useful work-related information and references in an accidental or unplanned way or by chance while looking for something else or while doing something else. Could you enlighten me about the frequency and importance of such accidental acquisition of information with examples ?

13. To probe into interdisciplinary nature and importance of 'browsing activity' : As part of your information collecting activities and keeping upto date in your field of specialisation you may browse/peruse through considerable sources of information related, loosely related, interdisciplinary and unrelated. Have you found any use or role played by information from sources which are interdisciplinary in nature which you initially thought as irrelevant

to your work and activity ? Similarly, have you found useful sources of information unrelated to your field of activity ? Could you elaborate by recalling an example ?

14. To know how far the publishing activity helps in collecting workrelated information :
How does your publishing activity (*i.e.*, writing papers for publication) help you to collect work-related information ?

15. To know how information is exchanged among coauthors :
You might have coauthored with many colleagues in ISRO as well as outside ISRO for publications. What are your views about exchange of work-related information among such group of coauthors ?

16. To check whether items/references cited are highly sought after information of the author :
You might have cited many references in your papers published in professional journals/periodicals or presented in professional conferences, seminars, etc. Do you consider that these references cited by you have/had a high place among your work-related information requirements ?

17. To forecast future requirements :
In view of the possible scientific and technological changes or consequential changes in the activities of your Organisation, do you foresee any trend or possible likely change in your future information requirements ?

18. General exploratory question :
Could you enlighten me on any other special features of your information-gathering habits, if any ?

Appendix 4

SECTION AND PROJECT AFFILIATION OF THE SPACE TECHNOLOGISTS

Section/Project/Division		Population		Response Population	
Code	Name	No.	%	No.	%
21	Control Systems	39	5.1	24	5.1
22	Sensors	41	5.4	33	7.0
75	EMD	22	2.9	14	3.0
50 & 64	Facilities	8	1.0	6	1.3
52	Workshop	47	6.1	20	4.2
53	Draughting	32	4.2	19	4.0
54 & 55	Central Electronic Fabrication Facility	35	4.6	18	3.8
56	Printed Circuit Board Laboratory	3	0.4	0	0.0
57	Hybrid Microcircuit Laboratory	2	0.3	1	0.2
58	Electronic Maintenance	8	1.0	6	1.3
59	Environmental Test Facility	27	3.5	17	3.6
60	Photo Laboratory	2	0.3	2	0.4
30	IGCTD	5	0.7	2	0.4
31,32 & 33	System Integration (Includes mechanical and electrical Integration)	40	5.2	25	5.3
35	Ground Checkout	31	4.1	24	5.1
71 & 74	IRS Project	21	2.7	15	3.2
40	MOPD	11	1.4	6	1.3
41	Mission Planning and Analysis	16	2.1	10	2.1
42	Orbital Mechanics	18	2.4	14	3.0
43	Data Processing & Computer Facility	26	3.4	10	2.1
10	MSD	1	0.1	1	0.2
11	Structures	33	4.3	19	4.0
12	Thermal Systems	33	4.3	14	3.0
13	Mechanisms	13	1.7	5	1.1
70	PPED	16	2.1	11	2.3
72	INSAT II	3	0.4	3	0.6

Section/Project/Division		Population		Response Population	
Code	Name	No.	%	No.	%
65	QA & TE	52	6.8	33	7.0
73	SROSS	16	2.1	8	1.7
69	TP	18	2.4	18	3.8
01	TTC & P	22	2.9	13	2.7
02	Digital Systems	38	5.0	24	5.1
03	VHF Microwave and Antenna	40	5.2	28	5.9
04	Power Systems	40	5.2	31	6.5
06	Communications	2	0.3	0	0.0
77 & 89	Others	3	0.4	1	0.2
	Total	765	100.0	475	100.5

Note : Forty two not known cases under population and 30 not answered and 30 anonymous cases under respone population are excluded.

Appendix 5

SPECIALISED FIELDS OF ACTIVITY OF THE RESPONDENTS

Code	Major Specialised Fields of Activity	No.
A1	Vacuum and Cryogenic Technology	3
A2	Optical systems; Electro-optical system; Optics	12
A3	Thin films; Thin Film coating	4
A4	Astronomy and astrophysics; Space astronomy; High energy astronomy	4
A5	Ground based systems	1
A6	X-ray and X-ray counter	1
A7	Detector system (Technical Physics)	1
B1	Remote Sensing	4
B2	Image processing; Pattern recognition; Computer Vision	3
C1	EMI/EMC (Electromagnetic Interference/ Electromagnetic Compatability)	2
C2	Test system; Test equipment; Testing; Evaluation; Calibration of Test equipment	19
C3	Sensors	2
C4	Instrument maintenance	3
C5	Mangnetometers (with respect to sensors)	1
C6	Instrumentation	5
C7	Ground support equipment related work	3
D1	Spacecraft control system (attitude control, AOCS, space guidance and control)	8
D2	Dynamic modelling of spacecraft	1
D3	Spacecraft dynamics	2
E1	Spacecraft structure; Structural design and analysis; Structural mechanics	11
E2	Vibrations; Vibration system; Vibration testing; Random vibrations; Acoustics	3
E3	Composite materials; Structural materials	4
E4	Polymers; plastics	1

E5	Holography; Holographic applications in structural testing	1
E6	Heat treatment of metals	1
E7	Stress analysis	1
F1	Inspection, screening and testing of active and passive components; Fabrication and testing of semiconductor devices	4
F2	Reliability and quality assurances; Failure analysis	12
G1	Power system; power electronics/energy devices	11
G2	Electro-chemistry	1
G3	Solar cells; Solar arays; Solar energy; Photo voltaics; Solar panels	8
H1	Integration of spacecraft	8
H2	Mechanical supports for spacecraft; Mechanics, Applied mechanics; Fixtures of spacecraft	7
H3	Thermal system; Heat transfer; Thermal control coating for spacecraft; Thermal projector	7
H4	Booms (deployable)	1
H5	Drawing, draughting and related	20
H6	Lubrication	2
H7	Tribology; Space tribology	3
H8	Workshop activities	4
H9	Balancing of spacecraft (with respect to vibrations)	1
HU	Fitting works	2
HV	Production engineering	1
HW	Mechanical Inspection	1
I1	Digital electronics; Digital circuits	6
I2	Telecommand system; Onboard telecommand system	2
I3	Telemetry	1
J1	Microcircuit; Hybrid microcircuit; Hybrid microcircuit packages	6
J2	PCB (Printed Circuit Board)	2
J3	Soldering (Electronic Components)	1
J4	Fabrication process control [of PCBs, HMCs, MIGs (Microwave Integrated Circuits) and other electronic and mechanical items]	2
J5	Wirings	1
J6	Electronic Lubrication; Electronic wiring	6
J7	LSI (Large Scale Integration) systems	1
J8	Amplifier; Preamplifier	1
K1	Microprocessor/microcomputer based system	8

L1	Communication engineering/technology; Satellite communication system	16
L2	Digital communication	5
L3	Transmitters	1
L4	Microwave communications; Microwaves; Microwave circuits; Microwave electronics; Antennas, VHF (Very High Frequency) systems	16
L5	Data reception and decodification	1
L6	Transponders	1
M1	Oribital mechanics, dynamics and analysis, attitude mechanics	6
M2	Space simulation	5
M3	Mission operations; Mission planning and analysis	5
M4	Ground system related activities	1
N1	Propulsion	1
Z1	Software development; Computer programming	30
Z2	Key punching	3
Z3	Data processing, Computer system related work	6
Z4	Computer communications	1
P1	Project management; Technical management	7
P2	Managerial planning and coordination; Planning and evaluation; Budgeting; Training	4
P3	Feasibility study, Project report; Preliminary studies; Spacecraft definition and design studies	2
P4	Spacecraft subsystems configuration analysis	1
P5	System analysis	2
P6	Coordination	1
Q1	Safety	1
R1	Airconditioning, Refrigeration; Diesel generator, Electrical maintenance	5
S1	Video production; photography; Reprography	4
	Total	356

EDUCATIONAL QUALIFICATIONS OF THE SPACE TECHNOLOGISTS

Code	Acronym	Qualification	No.
U01	COFT		1
U11	ITC	Industrial Training Course	3
U12	ITI	Industrial Training Institute Course	116
U14&	NAC	National Apprenticeship Certificate	
U17	NTC	National Trade Certificate	
U13 &	Inter	Intermediate/Pre-university course	3
U21	PUC		
U24 &	SSLC	Metriculation/Secondary School Leaving Certificate/	
U25	X Std.	X Standard	6
U26	NCTVT	National Council of Trade Vocational Training	5
U29		Fitter Certificate Course	1
U30		Key Punching (Certificate Course)	1
U31		Wireless Mechanic Grade I Certificate	1
U32		Radio Mechanic Certificate	2
U33		Telecommunication Technician Certificate	1
U34		Apprenticeship Training (Certificate Course)	1
U35		Certificate Course in Electronics	2
U36	IX Std	IX Standard	1
U38	VII Std	VII Standard	1
		Sub-total	145
D		Unspecified	2
D02	DCE	Diploma in Civil Engineering	3
D07	DORT		1
D08	DECE	Diploma/Licenoeate in Electronics &	
D23	LECE	Communication Engineering	5
D09	DEE	Diploma in Electronics Engineering/Electronics and/or	
D12	DERE	Radio Engineering/Electronics & Telecommunication	
D13	DETE	Engineering/Telecommunication Engineering	

D21	DTE		
D20	DRE		80
D16	DME	Diploma/Licenceate in Mechanical Engineering	
D25	LME		
D17	DMIT		54
D26	DCDM		3
D28	DFT	Diploma in Film Technology	1
D31		Diploma in Instrumentation & Control Engineering	1
D32		Diploma in Airconditioning/Refrigeration	1
		Sub-total	148
B		Unspecified	1
B01	AMIE	Associate Member of Institute of Engineers	9
B02	BA	Bachellor of Arts	2
B03	BBA/	Bachellor of Business Administration/	
	BBM	Management	6
B05	BE	Bachellor of Engineering	123
B07	B.Sc	Bachellor of Science	22
B08	B.Sc Engg.	Bachellor of Science (Engineering)	5
B09	B.Tech.	Bechellor of Technology	16
B12	GRADIETE		9
B13	B.E. (Hons)	Bachellor of Engineering (Honors)	3
		Sub-total	196
M02	MBA	Master of Business Administration	3
M05	ME	Master of Engineering	45
M06	MS	Master of Science	3
M07	M.Sc.	Master of Science	63
M08	M.Sc. Tech.	Master of Science (Technology)	10
M09	M.Tech.	Master of Technology	64
M10	M.Sc. Engg.	Master of Science (Engineering)	8
		Sub-total	195
P03	Ph.D	Doctor of Philosophy	32
		Not Known/Not Answered	120
		Grand-Total	837

Appendix 7

INTERCORRELATION OF PURPOSES OF SEEKING INFORMATION, TYPE OF INFORMATION SOUGHT AND DEPENDENCE ON TRADE LITERATURE (PEARSON PRODUCT MOMENT CORRELATION COEFFICIENTS)

Question Number		Question Number 2B	2C	1D	1E	3.1J
2B	Purpose of Designing and Development of Products and Components	—				
2C	Purpose of Setting up and use of equipment	0.40	—			
1D	Need for methods, process and procedures information	0.26	0.29	—		
IE	Need for product, material, equipment and operatus information	0.39	0.40	0.46	—	
3.1J	Dependence on Trade Literature	0.35	0.43	0.13*	0.44	

Key : *, Excepting this all other correlations are statitically significant at 0.05 significance level.

Appendix 8

ORGANISATIONAL AFFILIATION OF MOST-OFTEN CONTACTED PERSONS OUTSIDE ISAC

Sl. No.	Acronym	Name of the Organisation	No. of persons cited
		OTHER CENTRES, UNITS, PROJECTS OF ISRO/DOS	
1.	APSU	Auxiliary Propulsion System Unit, Bangalore	3
2.	CED	Civil Engineering Division, Bangalore	4
3.	DOS	Department of Space Office, Bangalore	7
4.	INSAT-I	Indian National Satellite System-I Project, Bangalore and Hassan	3
5.	ISROHQ	Indian Space Research Organisation Headquarters, Bangalore	7
6.	PRL	Physical Research Laboratory, Ahmedabad	1
7.	SAC	Space Application Centre, Ahmedabad	7
8.	SHAR	Shriharikota High Altitude Range	1
9.	VSSC	Vikram Sarabhai Space Centre, Trivandrum	3
			36
		ORGANISATIONS WITHIN THE COUNTRY	
		A. RESEARCH ORGANISATIONS	
1.	ADE	Aeronautical Development Establishment, Bangalore	1
2.	CEERI	Central Electronics Engineering Research Institute	1
3.	CMTI	Central Machine Tool Institute, Bangalore	6
4.	CPRI	Central Power Research Institute, Bangalore	2
5.	DRDO	Defence Research and Development Organisation	1
6.	GTRE	Gas Turbine Research Establishment, Bangalore	6
7.	LRDE	Electronics and Radar Development Establishment, Bangalore	2
8.	ONGC	Oil and Natural Gas Commission	1
9.	NAL	National Aeronautical Laboratory, Bangalore	10
10.	RRI	Raman Research Institute, Bangalore	2
11.	TIFR	Tata Institute of Fundamental Research (Bangalore + Bombay = 3 + 2)	5
			37

B. INDUSTRIES

12.	BEML	Bharat Earth Movers Limited	1
13.	HMT	Hindustan Machine Tools Limited, Bangalore	2
14.	BEL	Bharat Electronics Limited, Bangalore	9
15.	BHEL	Bharat Heavy Electricals Limited, Bangalore	3
16.	BRW	Bangalore Rubber Works	1
17.	BTS	Badarpur Tharmal Station, Delhi	1
18.	ELICO	ELICO Private Limited	1
19.	HIL	Hindustan Indane Limited, Bangalore	1
20.	ITI	Indian Telephone Industries, Bangalore	6
21.	MICO	Motor Industries Company Limited, Bangalore	1
22.	NGEF	New Government Electrical Factory, Bangalore	1
23.	SAIL	Steel Authority of India Limited, Bangalore	1
24.	HCL	Hindustan Computers Limited, Bangalore	1
25.	NGRI	National Geophysical Research Institute, Hyderabad	1
			30

C. BUSINESS FIRMS

26.	BLUES	Blue Star, Bangalore	3
27.	DATAM	Datamatics Corporation, Bangalore	1
28.	HINDT	Hinditrans, Bangalore	2
29.	INGR	Ingersoll-Rand, Bangalore	1
30.	IRTS	IR Technology Services, Bombay	1
31.	KELTR	Keltron, Bangalore	2
32.	L& T	Larson and Toubro	1
33.	MCM	Madura Coats, Madurai	1
34.	MICRO	Micronic Devices	1
35.	PSIDS	PSI Data Systems	1
36.	VIPIP	VIP India Limited, Poona	1
37.	Unnamed Business Firm, Madras		1
			16

D. ACADEMIC INSTITUTIONS

38.	BIT	Bangalore Institute of Technology	1
39.	BITP	Birla Institute of Technology & Science, Pilani	2
40.	BUB	Bangalore University, Bangalore	3
41.	IISB	Indian Institute of Science, Bangalore	47
42.	IITB	Indian Institute of Technology, Bombay	1
43.	IITD	Indian Institute of Technology, Delhi	1

44.	IITM	Indian Institute of Technology, Madras	8
45.	MEI	Mysore Education Institute	1
46.	MITM	Madras Institute of Technology, Madras	1
47.	NCT	National College, Trichi	1
48.	RCEB	Ramaiah College of Engineering, Bangalore	1
49.	SEC	Salem Engineering College	1
50.	SITT	Siddaganga Institute of Technology, Tumkur	1
51.	TCEM	Thiyagarja College of Engineering, Madurai	1
52.	UVCE	University Visveswaraiah College of Engineering, Bangalore	1
			71

E. OTHER ORGANISATION

53.	BT	Bangalore Telephones	1
54.	CWC	Central Water Commission, Bangalore	2
55.	KFIDC	Karnataka Fisheries Development Corporation, Bangalore	1
56.	P&T	Post and Telegraphs, Bangalore	2
57.	SRBTC		
58.	UIICL	United Indian Insurance Company Limited	1
			8

ORGANISATIONS OUTSIDE THE COUNTRY

1.	AEGT	Aircraft Engine Group, Telefunken	2
2.	CNES	Center National E' Eluder Spatiales, France	2
3.	CUC	Concordia University, Canada	1
4.	DFVLR	Deutsche Forschungs-Und Versachasantalt Fuer Luft-Und Roumfaht, West Germany	1
5.	DLRLH		1
6.		Dow Corning, USA	1
7.	ESA	European Space Agency, France	8
8.	ESTEC	European Space Research & Technology Centre (ESA), France	3
9.	GRGS		1
10.	JPL	Jet Propulsion Laboratory, (NASA) USA	2
11.		Leibold, USA	1
12.		Lutech Inc, USA	1
13.	MATRA	Material Para Tratamiento, Versailles, France	1
14.	NASA	National Aeronautics & Space Administration, USA	3
15.	NOAA	National Oceanic and Atmospheric Administration	1
16.	RAE	Royal Aeronautic Establishment	3

17.	S/C		1
18.	SAO	Smithsonian Astrophysical Observatory	1
19.		SOMEMEA, Toulose	1
20.		Spectro Laboratory, USA	2
21.	UOTO	University of Texas, Austrin	1
22.		Wacker Cheim	1
			38
		Total	236

Appendix 9

Publications, Lectures, Awards and Documents of the IST (Based on ISAC Annual Reports)

Year	Publications	Guest Lectures/ Presentations	Awards	Documents Prepared				Total
				Proposal/ Feasibility Report	Test Plan/ Test Results Reports	Technical Documents	Budget & Nontechnical Documents	
1978-79	49	NA	NA	9	59	96	16	180
1979-80	44	NA	NA	18	53	158	6	235
1980	41	33	NA	19	56	229	11	315
1981	33	13	5	18	37	257	5	317
1982	37	32	2	19	36	202	NA	257
1983	63	21	1	11	17	222	7	257
1984	49	7	5	11	34	193	11	249
Total	321	111	13	105	292	1357	56	1810

Key : NA, Not Available.

Appendix 10

Coauthorship Trend for the Organisation as a whole (Based on ISAC Annual Reports)

No. of authors	1978-79	1979-80	1980	1981	1982	1983	1984	Total	
								Papers	Authorships
01	16	18	12	5	12	23	21	107	107
02	7	10	11	11	13	15	13	80	160
03	9	7	8	11	4	12	11	62	186
04	5	6	2	4	4	4	2	27	108
05	0	1	1	2	1	3	2	10	50
06	2	1	0	0	2	4	0	9	54
07	4	1	1	0	1	0	0	7	49
08	2	0	0	0	0	2	0	4	32
09	2	0	0	0	0	0	0	2	18
10	1	0	0	0	0	0	0	1	10
et. al/ others	1	0	6	0	0	0	0	7	-
Not Known	0	0	0	0	0	5	0	5	-
Total	49	44	41	33	37	68	49	321	774

Note: Degree of collaboration is 0.66; excluding papers with et.al and not known authors, the multiplicity index is 2.51; and productivity index is 0.40

BIBLIOGRAPHY

"Is your lab well cited ?" *Nature* 227 (5255) 18 July, 1970 : 219.

"Science in India : excellence in the midst of poverty." *Nature* 308 (5960) 12 April 1984 : 581-584.

Ackoff, Russell L. et al, Designing a National Scientific and Technological Communication System (The SCATT Report). Pennsylvania : University of Pennsylvania Press, 1976.

Adam, Ralph, "Language and information retrieval in social sciences." *Aslib Proceedings* 34 (9) September 1982 : 394-405.

Aims, A., "Survey of information needs of physicists and chemists." *Journal of Documentation* 21 (2) June 1965 : 83-112.

Allen, R. H. et al., Study of the Airforce Maintenance Technical Data System. MRL-TDR-6285. Ohio : Wright Patterson Airforce Base, Behavioural Sciences Laboratory, 1962.

Allen, Thomas J. Utilization of Information Sources During R&D Proposal Preparation. MIT-97-64. Cambridge : MIT, Alfred P. Sloan School of Management, 1964.

— — Sources of Ideas and their Effectiveness in Parallel R&D Projects. MIT-130-65 ; PB-168430. Cambridge : MIT, Alfred P.Sloan School of Management, 1965.

— — "Studies of the problem Solving Process in Engineering design." *IEEE Transactions on Engineering Management EM-13* (2) June 1966 : 72-83.

— — Differential Performance of Information Channels in the Transfer of Technology. MIT Working Paper 196-66. Cambridge, Massachu setts : MIT, 1966.

— — "Sources of ideas in parallel R&D projects". In H.D.Learner ed. *Research Programme Effectiveness.* New York : Gordon & Breach, 1966.

— — "Performance of information channels on the transfer of technology." *Industrial Management Review* 8 (1) 1966 : 87-98.

— — "Organisational aspects of information flow in technology." *Aslib Proceedings* 20 (11).November 1968 : 433-454.

— — "Information needs and uses". In Carlos A. Cuadra and Ann W.Luke eds. Annual Review of Information Science and Technology. Chicago : *Encyclopedia Brittanica Inc.*, 1969. Vol. 4 ; 3-29

— — Meeting the Technical Information Needs of Research and Development Projects. MIT-284-67. Cambridge : MIT, Alfred P.Sloan

School of Management, 1969.

— — "Communication networks in R&D laboratories." *R&D Management 1* (1) October 1970 : 14-21.

— — "Roles in technical communication networks". In Carnot E.Nelson and Donald K. Pollock eds. *Communication among Scientists and Engineers*. Massachusetts : Heath Lexington Books, 1970. 191-208.

— — Technology Transfer to Developing Countries : The International Technological Gatekeeper. MIT-507-71. Cambridge : MIT, Alfred P. Sloan School of Management, 1971.

— — Managing the Flow of Technology : Technology Transfer and the Dissemination of Technological Information within the R&D Organisation. Cambridge :MIT Press, 1977.

Allen, Thomas J. et al. Time Allocation among three Technical Information Channels by R&D Engineers. MIT Working Paper No.184 66. Cambridge, Massachusetts : MIT, Alfred P. Sloan School of Management, 1966.

— — *Problem of Internal Consulting in Research and Development Organisation*. Washington, D.C. : NSF, 1969.

— — "International technological gatekeeper." *Technology Review* 73 (3) March 1971 : 37-43.

Allen, Thomas J. and Maurice P. Andmen Jr. Time Allocation Among Three Technical Information Channels by R&D Engineers. MIT-131-65. Cambridge : MIT, Alfred P.Sloan School of Management, 1965.

Allen, Thomas J. and Stephen I. Cohen. "Information flow in R&D laboratories." *Administrative Science Quarterly* 14 (1) January February 1970 : 12-19.

Allen, Thomas J. and A. R. Fustfeld. "Research laboratory architecture and the structuring of communications." *R&D Management* 5 (2) 1975 : 153-164.

Allen, Thomas J. and P. G. Gerstberger. Criteria for Selection of an Information Source. MIT-284-67. Cambridge : MIT, Alfred P.Sloan School of Management, 1967.

— — "Criteria used by research and development engineers in the selection of an information source." *Journal of Applied Psychology* 52 (4) 1968 : 272-279.

Allibone, T. E. "Scientist and his needs." *Bulletin of the Institute of Information Science* 4 (2) 1965 : 3-15.

Almeida, James A. and Richard L. Harvey. Analysis of the Relationships between Demographic Factors and Maintenance Technicians Attitude towardsTechnical Data. Master's Thesis ; AFIT-LSSR-5-79B

; AD-A076919/0.Ohio : An University,U.S.Airforce Institute of Technology, School of Systems and Logistics, 1979.

American Institute of Physics, Physics Information Division. Program for a National Information System for Physics and Astronomy 1971-75. New York : AIP, 1970.

———AIP Program for Physics Information : A National Information System for Physics and Astronomy 1972-76. New York : AIP, 1970.

— — American Psychological Association. Reports of the American PsychologicalAsssociation's Project on Scientific Information Exchange in Psychology. Washington : APA,1963.

— — Overview of the Structure, Objectives and Findings of the American Psychological Association's Project on Scientific Information Exchange in Psychology. APA-PSIEP-Overview Report B. Washington : APA, 1963.

Amey, G. X. Channel Hierarchies for Matching Information Sources to Users Needs. DSIS-Report-R2. Ottawa : Canadian Defense Research Board, 1969.

Amick, Daniel James. "Scientific elitism and the information system of science." *Journal of the American Society for Information Science* 24 (5) September-October 1973 : 317-327.

Andrews, Frank M. ed. "Scientific Productivity : The Effectivenessof Research Groups in Six Countries" Cambridge : Cambridge University Press, 1979.

Anthony, L. J. et al. "Growth of the literature of physics. *"Reports of Progress in Physics* 32, 1969 : 709-767.

Arndt, J. et al. "Information exchange among scientists : A two step sociometric study." *Sociology* 14 (3) August 1980 : 441-448.

Atanasiu, Pia and Liana Bogden. "Some considerations on the Questionnaires used in user studies." *Studies in Cerc Docum* 12 (3) September 1970.

Atanasiu, Pia and E. Toma. "Some methods employed in the study of the users' needs and requirements." *Studies in Cerc.Docum.* 12 (1) March 1970 : 15-27.

Atherton, Pauline. "Views of the communication network of scientific and technical information." *International Forum on Information and Documentation* 1 (1) 1975 : 10-12.

——"Research in information science : an assessment". In A.Debons and W.J. Cameron eds. *Perspectives in Information Science*. Leyden : Noordhoff, 1975. 665-679.

— — " *Handbook for Information Systems and Services.*" Paris : Unesco,

1977.

Atkin, Charles. "Instrumental utilities and information seeking". In Peter Clarke ed. *New Models for Mass Communication. Research*. Beverly Hills, California : Sage Publications, 1973. Vol. 2.

Auerbach Corporation. DoD User Needs Study, Phase I. Final Technical Report, 2 vols. DoD 1151-TR-3 ; AD 615 501,2. Philadelphia, pa.: Auerbach Corporation, 1965.

— — "What's wrong with information retrieval?" *Machine Design* 38 (16) 7 July 1966 : 106-109.

Auster, Ethel. "Organisational behaviour and information seeking lessons for librarians." *Special Libraries* 73 (3) July 1982 : 173-182.

Avramescu, A. "Modelling scientific information transfer." *International Forum on Information and Documentation* 1(1) 1975 : 13-19.

Ayres, F. H. et al. "Author versus titles : A comparative survey of the information which the user brings to the library catalogue." *Journal of Documentation* 24 (4) December 1968 : 266-272.

Bach, Harry. "Scientific literature use : A survey." *Special Libraries* 48 (9) 1957 : 466.

Back, Kurt H. "Behaviour of scientists : communication and creativity." *Sociological Inquiry* 32 (1) Winter 1962 : 82-87.

Baker, N. R. et al. "Effects of perceived needs and means on the generation of ideas for industrial R&D projects." *IEEE Transactions on Engineering Management* EM-14 (4) December 1967.

Baker, N. R. and J. R.Freeland. "Structuring information flow to enhance innovation." *Management Science* 19 (1) September 1972 : 105-116.

Baker, Sharon L. "Overload, browsers and selections." *Library and Information Science Research* 8 (4) October-December 1986 : 315-329.

Barber, A. S. "Critical review of the surveys of scientists' use of libraries". In W. L. Saunders ed. *The Provision and Use of Library and Documentation Services.* Oxford : Pergamon, 1966. 145-179.

Barber, Bernard. "Resistance by scientists to scientific discovery." *Science* 134 (3479) September 1961 : 596-602.

Bare, Carole E. "Conducting user requirement studies in special libraries." *Special Libraries* 57 (2) February 1966 : 103-106.

Barkey, P. "Patterns of student use of college library." *College and Research Libraries* 27 1966 : 277-282.

Barnes, R. C. M. "Some recent investigations into information use at

AERE, Harwell". In Looking Forward in Documentation : Papers and Discussions of 38th Aslib Annual Conference, September/ October 1964. London : Aslib, 1964. 4.1-4.5.

———"Information use studies : Pt. 2-Comparison of some recent surveys." *Journal of Documentation* 21 (3) September 1965 : 169-176.

Barnett, C. C. "Library surveys and user surveys". In M.A. Whatley ed. *British Librarianship and Information Science 1971-75*. London : The Library Association, 1977.

Bath University Library. Investigation into Information Requirements of the Social Sciences. Research Report 1-5. Bath : University of Bath Library, 1971.

———Comparison of results of science user studies with INFROSS. Bath : Bath University Library, 1971.

Bauer, Raymand A. "Audience". In Ithiel deSola Pool, et al. eds. *The Handbook of Communication*. Chicago : Rand McNally College Publishing Co., 1973. 141-152.

Beal, George M. et al. "*Knowledge generation, exchange and utilisation*. London : Westview Press, 1986.

Bebout, Lois et al. "User studies in the humanities : A survey and a proposal." *RQ* 15 (1) Fall 1975 : 40-45.

Bernal, J. D. "Preliminary analysis of pilot questionnaire on the use of scientific literature". In Report of the Royal Society Scientific Information Conference, 21 June-2 July 1948. London : Royal Society, 1948. 589-637.

———"Supply of information to the scientist : Some problems of the present day." *Journal of Documentation* 13 (2) December 1957 : 195-208.

———"Transmission of scientific information : A user's analysis". In International Conference on Science Information, Washington, D.C., 16-21 November 1958, 2 vols. Washington, D.C. : National Academy of Sciences, National Research Council, 1959.

———"Scientific information and its users." *Aslib Proceedings* 12 (12) December 1960 : 432-438.

Bernard, Jessie et al. Informal Communication among Bioscientists. Washington, D.C. : George Washington University, 1963/1964.

Borul, Lawrence and Allan Karson. "Evaluation of the Methodology of the DoD User Needs study". In Proceedings of the 31st Meeting and 1965 Congress of FID in Co-operation with ADI,7-16 October 1965. Washington,D.C. : Spartan Books, 1965. 2 : 151-157.

Bethell, John P. "Communications in an International Research Laboratory" London : The City University, Centre for Information Science,

1972.

Bitz, A. et al. "Production, Flow and Use of Information in Research Laboratories in different Sciences"Manchester : Manchester Business School, 1975.

Blagden, John. *Do We Really Need Libraries*? New York : K.G. Saur/ Clive Bingley, 1980.

Blaivas, Alex et al. "Consensuality of peer nominations among scientists." *Knowledge : Creation, Diffusion, Utilization* 4 (2) December 1982 : 252-270.

Blom, A. "Theoretical model for studying the information needs of scientists. Pt. 2." *South African Journal for Librarianship and Information Science* 49(2) October 1981 : 71-75.

——"Theoretical model for studying the information needs of scientists. Pt.1." *South African Journal for Librarianship and Information Science* 49 (1) July 1981 : 34-38.

Bodensteiner, W. D. Information Channel Utilization under varying Research and Development Project Conditions : An aspect of Interorganizational Communication Channel Usage. Ph.D. Thesis. Austin : University of Texas, 1970.

Boone, Morell Doaglas. Expectancies and Values as Predictors of Motivation of Pre-decisional Information Search. Ph.D. Thesis. New York : Syracuse University, 1980.

Bottle, R. T. "Scientists,information transfer and literature characteristics." *Journal of Documentation* 29 (3) September 1973 : 281-294.

Bozeman, B. "Public managers' acquisition of scientific and technical information." *R&D Management* 11 (1) 1981 : 33-35.

Brember, V. L. "Linking a medical user survey to management for effectiveness : II. Checklandsoft systems study." *Journal of Documentation* 4 (12) June 1985 : 59-74.

Briggs, W. G. Survey of MIT Science Library Users : (a) Patterns of User Behaviour, and (b) Effect of Circulation Time Upon Renewals. Cambridge : MIT, 1962.

Brittain, J. Michael. "Information and Its Users :A Review with Special Reference to Social Sciences" Bath :Bath University Press, 1970.

——User Studies in Education and the Feasibility of an International Survey of Information Needs in Education. Bath : Bath University, 1971.

——"Information and its Users -A Review with Special Reference to Social Sciences." New York : Wiley Inter-science, 1971.

—— "Information needs and applications of results of user studies". In Anthony Debons and Williams Cameron eds. *Perspectives in Infor-*

mation Science. Leyden : Noodhoff, 1975. 425-447.

— — "Pitfalls of user research and some neglected areas." *Social Science Information* Studies 2 (3) July 1982 : 139-148.

Brook, J. H. West. "Identifying significant research." *Science* 132, 1960 : 1229-1234.

Buel, W. D. "Biographical data and the identification of creative research personnel." *Journal of Applied Psycho-logy* 49 (3) 1965 : 318.

Bunch, Steven E. et al. Analysis of United States Air Force Maintenance Technical Data from the Maintenance Technician's Perspective. Master's Thesis ; AFIT-LSSR-11-78A. Ohio : Air University,U.S.Air Force Institute of Technology, School of Systems and Logistics, 1978.

Burton, R. E. and B. A. Green. "Technical reports in physics literature." *Physics Today,* 14 October 1961 : 35-37.

Bush, G. C. et al. "Attendance and use of the science library at MIT." *American Documentation* 7 (2) April 1956 : 87-109.

Butler, H. L. Enquiry into the statement of motives by readers Ph.D Thesis. Chicago : University of Chicago, 1939.

Butler, Meredith and Bonnie Gratech. "Planning a user study : Process defined." *College and Research Libraries* 43 (4) July 1982 : 320-330.

Buza, P. "Information gathering habits of technical experts and oral information." Tudomanyos Es Muszaki Tajekoztatas 18 August September 1971 : 629-654.

Campbell, David E. and Theodore, M. Shlechter. "Library design influences on user behaviour and satisfaction." *Library Quarterly* 49 (1) January 1979 : 26-41.

Case Institute of Technology, Operations Research Group. Study of the Dissemination and Use of Recorded Scientific Information Pt.I. *Journal Reading by Physicists and Chemists* Pt.II-*Cost of Journal Publication.* Pt. III - *Effect of condensation on Comprehension of Journal Articles.* Cleveland, Ohio : Case Institute of Technology, 1960.

Chafin, Roy L. "User characteristics for information system design". In Alan R.Benenfeld and Edward John Kazlauskas, eds. *Communicating Information - Proceedings of the 43rd ASIS Annual Meeting,1980,* Anahein, California, 5-10 October 1980. New York : *ASIS,* 1980. 17 : 317-319.

Chen, Ching-Chih. "How do scientists meet their information needs." *Special Libraries* 65 (7) July 1974 : 272-280

Clader, N. "What they Read and Why?" *DSIR Problems and Progress in Industry 4*. London : HMSO, 1959.

Clarke, Peter and Jim James. "Effects of situation, attitude intensity and personality on 'information seeking'." *Sociometry* 30 (3) September 1967 : 235-245.

——"Effects of situation, attitude intensity and personality on 'information seeking'." *Sociometry* 30 (3) September 1967 : 235-245.

Cole, P. F. "Analysis of reference questions records as a guide to the requirements of scientists." *Journal of Documentation* 14 (4) December 1958 : 197-207.

Cole, Stephen and Jonathan R. Cole. "Scientific output and recognition : A study in the operation of the reward system in science." *American Sociological Review* 32 (3) June 1967 : 377-390.

Collins, H. M. "TEA set : tacit knowledge and scientific networks." *Science Studies* 4 (4) 1974 : 165-186.

Columbia University, Bureau of Applied Social Research. Flow of Information among Scientists : Problems, Opportunities and Research Questions (Study Director :Herbert Menzel). New York :Columbia University, BASR, 1958.

——Review of Studies in the Flow of Information among Scientists (Study Director : Herbert Menzel) 2 vols. New York : Columbia University, BASR, 1960.

Compton, Bertita E. "Scientific communication". In Ethiel deSola Pool et al. eds. *Handbook of Communication*. Chicago : Rand McNally College Publishing Co., 1963. 755-778.

Compton, Bertita E. and William D. Garvey. "Information functions of an international meeting." *Science* 155 (3) March 1967 : 1648-1650.

Connolly, Terry. "Communication nets and uncertainty in R&D planning." *IEEE Transactions on Engineering Management, EM- 22* (2) May 1975 : 50-54.

Cooper, M. "Current information dissemination : Ideas and practices." *Journal of Chemical Documentation* 8 1968 : 207-218.

Cooper, William S. "Why bother theory of information usage." *Journal of Informatics* 2 (1) April 1978 : 2-5.

Coover, Robert W. "User needs and their effect on information centeradministration : a review 1953/66." *Special Libraries* 60, September1969 : 446-456.

Cott, Van. "National information system for psychologists : a proposed solution for a pressing problem." *American Psychologist*

25 (5) May 1970 : i-xx.

———*Collaboration, Communication and Influence : A Study of the Effects of Formal and Informal Collaboration among Scientists.* Baltimore Md. Johns Hopkins University, 1968.

———"Social structure in a group of scientists : Test of the 'invisible college' hypothesis." *American Sociological Review* 34 (3) 1969 : 335-352.

— — "Nature of scientific communication and influence." *International Social Science Journal* 22 (1) 1970 : 28-41.

———"Information needs and uses". In Carlos A. Cuadra ed. *Annual Review of Information Science and Technology.* Chicago : *Encyclopedia Britanica* Inc., 1971. Vol. 6 : 3-39.

— — "Invisible Colleges : Diffusion of Knowledge in Scientific Communication" Chicago : The University of Chicago Press, 1972.

Crawford, Susan. "Information needs and uses". In Martha E. Williams ed. *Annual Review of Information Science and Technology.* Martha E. Williams ed. New York : Knowledge Industries Inc. for ASIS, 1978. Vol. 13 : 61-81.

Crawford, Susan Young. "Communication centrality and performance". In *Proceedings of the American Society for Information Science,* Philadelphia, October 1970. Pennsylvania : ASIS, 1970. 45-48.

— — "Informal communication among scientists in sleep research." *Journal of American Society for Information Science* 22 (5) September-October 1971 : 301-310.

Cronin, Blaise. "Assessing user needs." *Aslib Proceedings* 33(2) February 1981 : 37-47.

— — "Invisible colleges and information transfer : a review and commentary with particular reference to the social sciences." *Journal of Documentation* 38 (3) September 1982 : 212-236.

Cronin, Blaise and M. Gudim. "Information and Productivity : A review of research." *International Journal of Information Management* 6 (2) June 1986 : 85.

Davis, Keith. "*Human Behaviour at Work : Organisational Behaviour*", 5th ed. New Delhi : Tata McGraw Hill, 1977.

Davis, R. A. "How engineers use literature." *Chemical Engineering Progress* 61 (3) March 1965 : 30-34.

Debons, Anthony ed. *Information Science : Search for Identity.* New York : Marcel Dekker, 1974.

Derr, Richard L. "Information seeking expressions of users." Journal of the *American Society for Information Science* 35 (2) March 1984 : 124-128.

Dervin, Brenda. "Everyday information needs of the average citizen : A taxanomy for analysis". In Manfred Kochen and Josephe Donhue eds. *Information for the Community.* Chicago : ALA, 1976. 19-38.

Dhawan, Sathish."Space and Industry." Sriram Memorial Lecture. N. p.n. p., 1983.

Disch, A. "Voice of the user : His information needs and requirements (which are not what the information specialist thinks they are)". In Problems of Optimization of User Benefit in Scientific and Technological Information Transfer : Panel Specialists' Meeting, Copenhagen, Denmark, 8-9 October 1975, AGARD-CP-179. Paris: AGARD, NATO, 1976. 14.1-14.8.

Dretske, Fred I. *Knowledge and the Flow of Information.* Oxford : Blackwell, 1981.

Dumanova, Aleksandra. "Study of the information needs of physicists (Rezultati ot anketa v Edinnirya tsentur po- fizika)." *Bibliotekar* 33 (5) May 1986 : 18-21.

Dux, W. "How do scientists inform themselves." *Boersenbl. Dt. Buchhandel, Leipzug.* 136 (25) 1969 : 495.

——"How scientists and engineers get their information : Results of an enquiry." *Informatik* 17 (4) 1970 : 46-50.

Egan, Margaret and Herman H. Henkle. "Ways and means in which research workers, executives and others use information". In J.H. Shera, et.al. eds. *Documentation in Action.* New York : Reinhold, 1956. Chapter VII, 137-159.

European Industrial Research Management Association. Information Requirements for Research and Development. EIRMA Working Group Report No. 8. Paris : EIRMA, 1973.

Evan, William. "Role strain and the norm of reciprocity in research organisations." *American Journal of Sociology* 68 (3) November 1962 : 346-354.

Evans, G. Edward. *Management Techniques for Librarians.* New York : Academic Press, 1976.

Evans, M. A. et al. "Communication patterns in two U.K. R&D laboratories." *R&D Management* 4 (3) 1974 : 141-147.

Fabisoff, S. G. and D. P. Ely. "Information and information needs." *Information Reports and Bibliographies* 5 (5) 1975 : 2-16.

Farms, G. F. "Effect of individual roles on performance in innovative groups." *R&D Management* 3 (1) January 1972 :23-28.

Fearn, Robin A. C. and Jessica S. Melton."Information centre user study". In Jeanne B.North ed. Proceedings of the 32nd Annual Meeting of

ASIS, San Francisco, California, 1-4 October 1969. Westport, Connecticut : Greenwood Publishers, 1969. 6 : 463-470.

Feinman, Stephen et al."Conceptual framework for information flow studies". In *Information Policies : Proceedings of the 39th Annual Meeting of the American Society for Information Science,* 4-9 October, Sanfrancisco, California Washington,D.C. : ASIS, 1976. 13,Pt.1. : 106.

Fine, Sara. "Research and the psychology of information use." *Library Trends* 32 (4) Spring 1984 : 441-460.

Fischer, William A. "Scientific and technical information and the performance of R&D groups". In Burton V. Dean and Joel L.Girdhar eds. *Management of Research and Innovation.* Amsterdam : North Holland, 1980. 67-89.

Fishenden, R. H. "Methods by which research workers find information". *In Proceedings of the International Conference on Scientific Information,* Washington, D.C., 16-21 November 1958. Washington, D.C. : National Academy of Sciences, National Research Council, 1959. 1 :163-179 (Preprints 1 p. 153-170).

Fishenden, R. M. "Information use studies, part I - past results and future needs." *Journal ofDocumentation* 21 (3) September 1965 : 163-168.

Fisher, W. A. and B.Rosen. "Search for the latent information star." *R&D Management* 12 (2) January 1982 : 61-66.

Fjallbrant, N. *Study of User Behaviour and Needs at Chalmers University of Technology Library,* Gottenberg, Sweden. CTHB Publication No.10. Gottenberg : Chalmers University of Technology Library, 1976.

Ford, Geoffrey. "Research in user behaviour in university libraries." *Journal of Documentation* 29 (1) March 1973 : 85-106.

Ford, Geoffrey ed. "User Studies : An Introductory Guide and Select Bibliography" CRUS Occasional Paper 1. Sheffield : University of Sheffield, CRUS, 1977.

Ford, N. "Relating information needs to learners characteristics in higer education." *Journal of Documentation* 36 (2) June 1980 . 99 114

Foskett, D. J. " *Pathways for Communications : Books and Libraries in the Information* Age" London : Clive Bingley, 1984.

Fox, Mary Frank. "Publication productivity among scientists : a critical review." *Social Studies of Science* 13 (2) May 1983 : 285-305.

Frame, J. D. and M. P. Carpenter. "International research collaboration." *Social Studies of Science* 9 (9) 1979 . 481-497.

Freeman, J. E. and A. H. Rubenstein. "Users and uses of scientific and technical information : Critical research needs." *Information Reports and Bibliographies* 4 (1) 1975 : 1-35.

Friendlander, Janet. "Clinician search for information." Journal of the *American Society for Information Science* 24 (1) January-February 1973 : 65-69.

Frohman, Alan. Determinants of Library Use in an Industrial Firm. Term Paper. Cambridge, Massachusetts : MIT, Alfred P. Sloan School of Management, 1969.

Frost, P. and R. Whitley. "Communication patterns in a research laboratory." *R&D Management* 1(2) 1971 : 21-79.

Fussler, H. H. "Characteristics of the research literature used by chemists and physicists in the United States." *Library Quarterly* 19 1949 : 119 143.

Gaffrey, Inez M. "Users, uses and supplies of STI services." Canadian *Journal of Information Science* 1(1) 1976 : 35-42.

Garg, Kailash Chandra and Ashok Kumar. "Information gathering of R&D scientists : A survey." *Annals of Library Science and Documentation* 31 (1-2) March-June 1984 : 63-73.

Garvey, William D. "Information exchange associated with national scientific meetings in relation to the general process of communication in science". In *Role of National Meeting in Scientific and Technical Communication*. Baltimore : Johns Hopkins University, 1970. Vol. 1.

— — "Dynamic scientific information user ".In Anthony Debons and William J. Cameron eds. *Perspectives in Information Science*. Leyden : Noodhoff, 1975. 483-509.

— — *Communication : The Essence of Science* Oxford : Pergamon, 1979.

Garvey, William D. et al. "Some comparisons of communication activities in the physical and social sciences". In C.E. Nelson and D.K. Pollock eds. *Communication among Scientists and Engineers. Massachusetts* : Heath Lexington Books, 1970. 61-84.

— — "Research studies in patterns of scientific communication : I, General description of research program." *Information Storage and Retrieval* 8 (3) June 1972 : 111-122.

— — "Research studies in patterns of scientific communication : III, Information exchange process associated with the production of Journal articles." *Information Storage and Retrieval* 8 (5) October 1972 : 207-221.

— — "Research studies in scientific communication : IV, The continuity of dissemination of information by 'productive scientists'." *Informa-*

tion Storage and Retrieval 8 (6) December 1972 : 265-276.

— — "Scientific information user".In Anthony Debons and William J.Cameron eds. *Perspectives in Information Science.* Leyden : Noodhoff, 1975.

Garvey, William D. and Belver C. Griffith. "Effect of convention presentations on information exchange behaviour and subsequent research". In Proceedings of the 27th Annual Meeting of the ADI. Washington,D.C. : ADI, 1964. 1 :201-213.

— — "Scientific communication as a social system." *Science* 157 (3792) September 1967 : 1011-1016.

Garvey, William D. and Kazvo Tomita. "Research studies in patterns of scientific communication : II, the role of the national meeting in society and technical communication." *Information Storage and Retrieval* 8 (4) August 1972 : 159-169.

Gaston, Jerry. *Big Science in Britain : A Sociological Study of the High Energy Physics Community.* Ph.D Thesis. Connecticut : Yale University, 1969.

— — "Reward system in British science." *American Sociological Review* 35 (4) August 1970 : 718-732.

— — "Communication and the reward system of science : A study of a national 'invisible college'." *Science Studies* 3 1973 : 245-273.

Gerstberger, Peter G. Investigation of the Criteria Used in Information Channel Selection by R&D Engineers. Master's Thesis. Cambridge: MIT, Alfred P. Sloan School of Management, 1967.

— — Preservation and Transfer of Technology in Researcn and Development *Organisations.* Ph.D Thesis. Cambridge, Massachusetts : MIT, Alfred P.Sloan School of Management, 1971.

Gerstberger, Peter G. and Thomas J. Allen. "Criteria used by research and development engineers in the selection of an information source." *Journal of Applied Psychology* 52 (4) August 1968 : 272-279.

Gilmore, John S.et al. Channels of Technology Acquisition in Commercial Firms and the NASA Dissemination Program. NASA-CR-790. Washington, D.C : Denver Research Institute, University of Denver, Colarado for NASA, 1967.

Glass, B. and S.H. Norwood."How scientists actually learn of work important to them". In *Proceedings of the international Conference of Scientific Information,* Washington,D.C.,16-21 November 1958, Washington,D,C. : National Academy of Sciences, National Research Council, 1959. 1 : 195-197(Preprints 1 p. 185-188).

Glock, C. and Herbert Menzel. "Flow of Information among Scientists :

Problems, Opportunities and Research Questions. "New York : Columbia University, BASR, 1958.

Goldberg, L.C. et al. "Local-cosmopolitan : Unidimensional or multidi mensional?" *American Journal of Sociology* 70 (6) May 1965 : 704-710.

Gooding, Richard Z. and John A. Wagner III. "A meta-analysis review of the relationship between size and performance : The productivity and efficiency of organisations and their subunits." A*dministrative Science Quarterly* 30(4) December 1985 : 462-481.

Goodman, A. F. Flow of Scientific and Technical Information : The Result of a Recent Major Investigation. Douglas Paper No.4516; AD 657 558. California : Douglas Aircraft Company, 1967.

Gordon, Michael D. "Critical reassessment of inferred relations multiple authorship scientific collaboration, the production of papers and their acceptance for publication." *Scientometrics* 2 (3) 1980 : 193-201.

Graham, W. R. et al. Exploration of Oral/Informal Technical Commu nications Behaviour. AD-669586. Washington, D.C. : American Institute for Research, 1967.

Gralewska-Vickery, A. "Communication and information needs of earth science engineers." *Information Processing and Management* 12 (4) 1976 : 251-282.

Gralewska-Vickery, A and H. Roscoe. Earth Science Engineers : Commnication and Information Needs. OSTI Report No. 5226. London : Imperial college, 1975.

Gray, Dwight E. "Physics abstracting." *American Journal of Physics* 19 (7) October 1950 : 417-424.

Gray, John and Brian Perry. *Scientific Information*, New York : Oxford University Press, 1975.

Great Britain Advisory Council on Scientific Policy. "Survey of information needs of physicists and chemists." *Journal of Documentation* 21 June 1965 : 83-112.

Griffith, Belver C. et al. "Informal contacts in science : A probabilistic model for communication processes." *Science* 173 (4) 9 July 1971 : 164-166.

Griffith, Belver C. and William D. Garvey. "Systems in scientific information exchange and the effects of innovation and change". In *Proceedings of the 27th Annual Meeting of ADI on Parameters of Information Science,* Philadelphia, Pennsylvania, 5-8 October 1964. Washington, D.C. : ADI, 1964. 1 : 194-195.

Griffith, Belver C. and A. J. Miller. "Networks of informal communication among scientifically productive scientists". In C.E. Nelson and D.K. Pollock eds. *Communication Scientists and Engineers.* Massachusetts : Heath Lexington Books, 1970. 125-140.

Grose, Daphne. "Some deprived information users." *Aslib Proceedings* 26 (1) January 1974 : 9-27.

Guilford, J. P. *Fundamental Statistics in Psychology and Education.* New York : McGraw Hill, 1956.

Haag, D. E. "Barriers limiting the usefulness of published information in the research environment." *Special Libraries* 75 (3) July 1984 : 214-220.

Hagstrom, Warren O. *Scientific Community* New York : Base Books Inc., 1965.

— — "Factors related to the use of different modes of publishing research in four scientific fields". In Carnot Nelson and Donald K Pollock eds. *Communication among Scientists and Engineers.* Massachusetts : Heath Lexington Books, 1970. 85-124.

Halbert, Michael H. and Russel L. Ackoff. "Operations research study of the dissemination of scientific information". In *Proceedings of the International Cconference on Scientific Information,* Washington, D.C., 16-21 November 1958. Washington, D.C. : Nationa! Academy of Sciences : National Research Council, 1959. 1 : 97-130 (Preprints 1 p.87-120).

Hall, Kevin R. and Eric Ritchie. "Study of communication behaviour in an R&D laboratory." *R&D Management* 5 (3) July 1975 : 243-245.

Hall, R. M. S. "Information for design". A paper presented at Engineering Design Show Conference, Brighton, 1969. N. p. n. p., 1969.

Hall, Robert W. "Investigation into the Information Habits of Scientists and Engineers in Industry." Bloomington : Indiana University, 1969.

— — Personal Effectiveness in Industrial Research. PhD Thesis. Indiana : Indiana University, Graduate School of Businees, 1972.

— — "Technical information habits of engineers." *Chemical Engineering Progress* 69 (3) March 1973 : 67-71.

Halperin, Michael R. and Alok K. Chakraborti. "Firm and industry characterstics influencing publications of scientists in large American companies." *R&D Management* 17 (3) July 1987 : 167-173.

Hanson, C. W. "Deciding what to read to keep informed." *Engineering* 190 (9) September 1960 : 378-379.

— — "Research on users' needs : where it is getting us?" *Aslib Proceedings* 16 (2) February 1964 : 64-78.

— — Review of Research and Experimentation on Engineering Information Particularly in Mechanical Engineering. OSTI Report No. 5202. 1974.

Harris, Colin. "User education and user studies". In L.J. Taylor ed. *British Librarianship and Information work 1976-1980.* London : The Library Association, 1982. vol. 2 : 145-164.

— — "User needs and user education". A paper presented at Fifth Information Officers Training Course (INFOTRAC)6-31 May 1985, Manila, Philippines. Organised by TECHNONET ASIA and APCTT National and International Transfer and Development. N. p. n. p., 1985.

Harris,T. W. "Contribution of research :information transfer and the adoption of technological innovation". In *Accelerating Innovation : Papers given at a Symposium held at the University of Nottingham, March1969. Organised by Aslib, U.K.Ministry of Technology and The Research Associations. London : Aslib, 1970. 11-15.

Hartley, Peter. "Communication studies at Sheffield city polytechnic". In Teaching Communication Skills to Engineers and Scientists in Higher Education - Proceedings of a One Day Conference, 17 November 1978, UWIST, Cardiff. Cardiff : UWIST, 1978. 243-248.

Hatt, Frank. "*Reading Process :A Framework for Analysis and Description"* London : Clive Bingley, 1976.

Havelock, Ronald G. "Research on utilization of knowledge". In Manfred Kochen ed. *Information for Action : From Knowledge to Wisdom.* New York : Academic Press, 1975. Chapter 7, 87-107.

— — "Research utilization". In Manfred Kochen and Joseph C. Donohue eds. Information for the Community. Chicago : ALA, 1976. Chapter 13,200-213.

Havelock, Ronald G. et al. "Planning for Innovation Through Dissemination and Utilization of Knowledge" Michigan : The University of Michigan, The Institute for Social Research,CRUSK, 1969.

Heenan, William F. and David C.Weeks. Informal Communication among Scientists : A Study of the Information Exchange Group Program. Final Report Pt.I. Washington, D.C. : Georgre Washington University, 1971.

Herner, Saul. "Information-gathering habits of workers in pure and applied sicence (how scientists seek and obtain information)." *Industrial and Engineering Chemistry* 46 (1) January 1954 : 228-236.

— — Determining Requirements for Atomic Energy Information from Reference Questions. ICST, 1958.

Herner, Saul et al. Evaluation of the Goddard Space Flight Centre Library.

NASA-CR-159969.Washington,D.C. :Herner & Co., 1979.

Herner, Saul and Mary Herner. "Information needs and uses in science and technology". In Carlos A.Cuadra ed. *Annual Review of Information Science and Technology.* New York : Interscience Publishers, 1967. Vol. 2 : 1-34.

Herner, Saul and N. Herner. "Determining requirements for atomic-energy information from reference questions". *In Proceedings of the International Conference on Scientific Information,* Washington, D. C. 16-21 November 1958. Washington,D.C. : National Academy of Sciences, 1959, 1 : 181-187.

Herner, Saul and Dewitt O. Myatt. "Building a functional library." *Chemical and Engineering News* 32 1954 :4980.

Hertz, D. B. and A.H. Robenstein. "Team Reasearch" New York : Columbia University, 1953.

Hills, Philip J. *Scientific Temperaments : Three Lives in Contemporary Science.* New York :Simon and Schuster, 1982.

Hintz, C. W. E. *Internationalism and Scholarship : A Comparative Study of the Reference Literature Used by American, British, French and German.* Ph.D Thesis. University of Chicago, 1952.

Hiscock, Jane E. "Does library usage affect academic performance ? A study of the relationship between academic performance and usage of libraries at the Underdale site of the South Australian College of Advanced Education." *Australian Academic and Research Libraries* 17 (4) December 1986 : 207-214.

Hoadley, Irene Braden and Alice S. Clark eds. "Quantitative Methods in Librarianship : Standards, Research, Management—Proceedings and Papers of an Institute held at the Ohio State University,3-16 August 1969." Contributions in Librarianship and Information Science No.4. Connecticut : Greenwood Press, 1972.

Hogg, I. H. and J.R. Smith. "Survey of the communication of information in the research and development branch of the United Kingdom Atomic Energy Authority". In *Proceedings of the International Conference on Scientific Information,* Washington,D.C , 16-21 November 1958. Washington, D.C. : National Academy of Sciences, National Research Council, 1959. 1 : 131-162 (Preprints 1 p.235 252).

Holland, Winford E. Intra and Interorganisational Communications Behaviour of Scientists and Engineers with High Information Potential. Doctoral Dissertation. Austin : University of Texas, 1970.

— — "Characteristics of individuals with high information potential in government research and development organisations." *IEEE*

Transactions on Engineering Management EM-19 (2) May 1972 : 38-44.

———"Information potential : a concept of the importance of information sources in a research and development environment." *Journal of Communication* 22 (2/3) June 1972.

———"Special communicator and his behaviour in research organisations: a key to the management of informal technical information flow." *IEEE Transactions on Professional Communication* PC-17 (3-4) September- December 1974 : 48-53.

Hounsell, Dai. "Relations between user and sources of information." *Education Libraries Bulletin* 25 (3) Autumn 1982 : 30-33.

India, Department of Science and Technology. Research and Development Statistics 1980-81. New Delhi : India, DST, 1982.

———Research and Development Funding Schemes of Central Government Agencies :General Information. New Delhi : India, DST, 1985.

International Federation for Documentation, Study Committee on Research on Theoretical Basis of Information. *Problems of Information User Needs* A.I. Mikhailov. Moscow : All Union Institute for Scientific and Technical Information, 1973.

Jahoda, G. "Information needs of science and technology - background review". In *Proceedings* of the 31st Meeting and 1965 Congress of FID, Washington, D.C., 7-16 October 1965. Washington : Spartan Books, 1966. 2 : 137-142.

— — "Information gathering and use habits of chemists." *Journal of Chemical Documentation* 9 (5) 1969 :153.

Jarvelin, Kalervo and Aatto J. Repo. "Knowledge work augmentation and human information seeking." *Journal of Information Science* 5 (2/3) November 1982 : 79-86.

Johns Hopkins University, Center for Research in Scientific Communi cation. Comparison of the Dissemination of Scientific and Technical Information, Informal Interaction, and the Impact of Information Associated with Two Meetings of the American Institute of Aeronautics and Astronautics. JHU-CRSC-Report No.1. Maryland : JHU, CRSC, 1967.

———Dissemination of Scientific Information, Informal Interaction, and the Impact of Information Received from Two Meetings of the Optical Society of America. JHU-CRSC-Report No.3. Maryland : JHU, CRSC, 1967.

———Dissemination of Scientific Information, Informal Interaction and the Impact of Information Associated with the 48th Annual Meeting of the American Geophysical Union. JHU-CRSC-Report No. 5. Maryland : JHU, CRSC, 1967.

Jhonston, Ron and Michael Gibbons. "Characteristics of information usage in technological innovation." *IEEE Transactions on Engineering Management* EM-22 (1) February 1975 : 27-34.

Jones, Pat. Some Social Aspects of Information Transfer. Research Report No. 44. London : The Polytechnic of North London, School of Library, 1981.

Kastens, Merrit L. " Information needs of applied research".. In *Modern Trends in Documentation.* New York : Pergamon Press, 1959. 23-30.

Katz, Elihu. "Two step flow of communication : an update report on hypothesis." *Public Opinion Quarterly* 21 (1) Spring 1957 : 61-68.

——– "Two step flow of communication : An update report on hypothesis". In Schramm Wilbur ed. *Mass Communications.* Ilinois : University of Illinios Press, 1960. 346-365.

Katz, Elihu and Paul F. Lazarsfeld. "Personal Influence : The Part Played by People in the Flow of Mass Communications" "Illinois : The Free Press, 1955.

Katz, Ralph. Does gatekeeping make a difference ?" Chemtech 13 (3) March 1982.

Katz, Ralph and M. Tushman. "Investigation into the managerial roles and career paths of gatekeepers and project supervisors in a major R&D facility." R&D Management 11 (3) 1981 : 103-110.

Kaufman, Harold, Factors related to Use of Scientific and Technical Information and Effectiveness of Engineers : A Longitudinal Study. DSI-77-18048, New York : Polytechnic Institute of New York.

Keenan, Stella and Margaret Slater. "Current awareness needs of physicists: Results of an Anglo-American study." *Journal of Documentation* 24 (2) June 1968 : 98-106.

Kelly, Alison. *The Missing Half : Girls and Science Education* Manchester: Manchester University Press, 1981.

Kelly, Kevin. "Information as a communicable disease : Infected by ideas." *Computers Today* 1 (11) January 1986 : 45-48.

Kepm, D A. *"Nature of knowledge : An Introduction for Librarians."* London: Clive Bingley, 1976.

Kenney, Larainc. "Implications of the needs of users of the design of a catalogue : A survey at the International Labour Office." *Journal of Documentation* 22 (3) September 1966 : 200.

Kent, Allen et al. "Use of Library Materials : The University of Pittsburgh Study" New York : Marcel Dekker, 1979.

Kidd, J.S. et al. "Information seeking behaviour of scientists as a guide to information product design". *In National Science Foundation, Division of Science Information.* Current Research on Scientific and Technical Information Transfer. New York : Jeffrey Norton Publishers Inc., 1977.

King, Donald W. and Vernon Eugene Palmour. "User behaviour : State-of-the-art paper". In C. Fenichel ed. Changing pattern in Information Retrieval : 10th ASIS Annual National Colloquium, Philadelphia, 3-4 May 1973. Washington, D.C. : ASIS 1974, 7-33.

—— "How needs are generated : What we have found out about them". In Nation Wide Provision and Use of Information : Aslib / IIS / LA Joint Conference, 15-19 September 1980, Sheffield - Proceedings. London : The Library Association, 1981. 68-79.

Knox, William T. "System for technological information transfer." Science 181 (4098) 3 August 1973 : 415-419.

Koch, T. "How do scientists inform themselves on the important literature : Investigated on the example of the faculty for the science of forestry of the Technical University Dresden in Tharandt." *Zbal. Biblwesen, Leipzig* 82 (11) November 1968 : 651-652.

Kochen, Manfred. "What makes a citizen information system used and useful". In Manfred Kochen and Joseph C. Donhue eds. *Information for the Community*, Chicago : American Library Association, 1976. Chapter 10, 149-170.

Kochen, Manfred ed. "Growth of Knowledge : Readings on Organization and Retrieval of Information". New York : John Wiley, 1967.

Koestler, Arthur. "Act of Creation. "New York : Macmillan, 1964.

Kolot, A. B. and V. R. Maksimov. "Matematicheskaya model analiza informatsionnykh potrebnostei po alternative nomupizanaku (Math ematical model of information needs analysis based on an alternative feature)." *Nauchno-Tekhnicheskaya Informatsiya Series* 2 (6) 1978: 8-11.

Kotani, Masao. "Communication among Japanese scientists domestically and with their counterparts abroad." *American Documentation* 13 (3) July 1962 : 320-327.

Kransberg, M. "Formal versus informal communication among researchers". In *Current Research on Scientific and Technical Information Transfer.* Jeffrey : Nation Publishers, 1977.

Kremer, Jeannette M. "Avaliacao de fontes de informacao usadas por engenheiros (Evaluation of information sources used by engineers)." *Revista de Bibliot-economua de Brasilia* 10 (2) July-December 1982

: 65-78.

Krikelas, James. "Information-seeking behaviour : Patterns and concepts." *Drexel Library Quarterly* 19 (2) Spring 1983 : 5-20.

Krummel, D.W. ed. "Introduction to trends in the scholarly use of library resources." *Library Trends* 25(4) 1977 : 725-732.

Kunz, Werner et al. "Methods of Analysis and Evaluation of Information Needs : A Critical Review" Munchen : Verlag Dokumentation, 1977.

Lamb, Gertrude H. "Rediscovery of librarianship". In Irene Braden Hoadley and Alice S. Clark eds. *Quantitative Methods in Librarianship : Standards, Research, Management :* Proceedings and Papers of an Institute held at the Ohio State University, 3-16, August 1969. Connecticut : Greenwood Press, 1972. 113-118.

Lamy, Alexus B. "Checking and measuring the level of service : Report of a survey undertaken by a major company into user needs". In Heather Taylor ed. *Information Management and Organisational Change-* Proceedings of Aslib Conference. London, 6-8 April 1981. London: Aslib, 1981. 10-26.

Lancaster, F. Wilfrid. "Assessment of the technical information requirements of users". In Alan Rees ed. *Contemporary Problems in Technical Library and Information Center Management - A State-of the-art.* Washington : ASIS, 1974. 59-85.

Lancaster, F. Wilfrid and Linda C. Smith. "Science, scholarship and the communication of knowledge." *Library Trends* 27 (3) Winter 1979: 367-388.

Landau, Robert et al (Working Group V). "User needs". In Anthony Debons and William J. Cameron eds. *Perspectives in Information Science.* Leyden : Noodhoff, 1975.

Langrish, L. et al. *Wealth from knowledge* London : Macmillan, 1972.

Leggett, Robert G. *"Do Engineers Read (or Buy) Books?."* Houston : Scholarly Publishing, 1976.

Leupost, M. "Information need, its nature and its implementation." *International Forum on Information and Documentation* 8 (3) 1983 : 3-7.

Levine, Emil H. "A generalized methodology for determining the correlation of the user requirements to the information system". In *Management of Information System*-Proceedings of 7th Mid Year Meeting of ASIS. Dallas Texas : Xerox Corporation, 1978.

Levine, J.M. et al. "Information seeking with limitations on available information and resources." *Human Factors* 17 (5) 1975: 502 513.

Levine, J.M. and M.G. Samet. "Information seeking with multiple sources of conflicting and unreliable information." *Human Factors* 15 (4)

1973 : 407-419.

Lewin, Kurt. "Channels of group life." *Human Relationships* 1 (2) 1947 143-153.

Ley, P. *"Quantitative Aspects of Psychological Assessment : An Introduction."* London : Duckworth, 1972.

Libby, Miles and Gerald Zaltman. Role and Distribution of Written Informal Communication in Theoretical High Energy Physics. AIP/SDD-1; USAEC-NYO-3732-1. New York : AIP, 1967.

Lickert, Rensis. "Motivational approach to organisations". In M. Haire ed. *Modern Organisational Theory.* New York : Wiley, 1959.

Lin, Nan. "Comparison between the scientific communication model and the mass communication model". In *Conference on Psychology of Technical Communication,* Philadelphia, Pennsylvania, 14-15 February 1972. New York : IEEE, 1972.

Lin, Nan et al. "A study of the communication structure in science". In C.E. Nelson and D.K. Pollock eds. *Communication among Scientists and Engineers.* Massachusetts : Health Lexington Books, 1970: 23-60.

Lin, Nan and William D. Garvey. "Information needs and uses". In Carlos A Cuadra ed. *Annual Review of Information Science and Technology.* Chicago : Encyclopedia Britanica Inc., 1972. Vol. 7 : 5-37.

Line, Maurice B. "Information uses and needs of social scientists : an overview of INFROSS." *Aslib Proceedings* 23 (8) August 1971 : 412-434.

——"Fortyseventh Aslib annual conference summing up : the information service in practice." *Aslib Proceedings* 26 (1) January 1974 :47-53.

——"Draft definition : Information and library needs, wants, demands and uses." *Aslib Proceedings* 26 February 1974 : 87.

——"Ignoring the user : How, when and why ?" In Nationwide Provision and Use of Information - *Proceedings* of ASLIB/IIS/LA Joint Conference 15-19 September 1980, Sheffield, *Proceedings.* London: The Library Association, 1981. 80-88.

Line, Maurice B. and Mavis Tidmarsh. "Student attitudes to the university library : a second survey at the Southampton University." *Journal of Documentation* 22 (2) June 1966 : 128.

Lingwood, David A. and Emile G. McAnany. "Scientific Information Flow and National Development : Study of Brazilian Chemists "Michigan: The University of Michigan, Institute for Social Research, CRUSK, 1971.

Lipetz, Ben-Ami. "Information needs and uses". In Carlos A. Cuadra ed. *Annual Review of Information Science and Technology.* Chicago :

Encyclopedia Britanica Inc., 1970. Vol. 5 : 3-32.

Lipetz, Ben-Ami and Peter Stangl. "User clues in initiating searches in a large library catalog". *In Proceedings of ASIS Annual Meeting*, Columbus, Ohio, 20-24, October 1968. New York : Greenwood, 1968. 5 : 137-139.

Longuet-Higgins, H.C. "Language of science." *Times Literary Supplement* (UK) (3558) 7 May 1970 : 505-506.

Lor, P.J. Information Needs of the General Practitioner : A Theoretical Model Based on an analysis of the General Practitioner's Cognitive Environment. Mousaion 11.8. Pretoria : University of South Africa, 1979.

Lubans, J. "On non-use of an academic library : A report of findings". In *Use, Mis-use and Nonuse of Academic Libraries.* New York : New York Library Association, College and University Libraries Section, 1970. 47-70.

Lufkin, J.M. and E. H. Miller. "Reading habits for engineers : a preliminary survey." *IEEE Transactions on Education* E-9 (4) December 1966 : 179-182.

Macadam, E.J. et al. "Technical Information of Engineers : What they Need and What They Get." London : Institution of Mechanical Engineers, 1969.

Maizell, Robert Edward. "The most creative chemist reads more." Industrial and Engineering Chemistry 50 October 1958 : 64A-65A.

— — "Information gathering patterns and creativity : A study of research chemists in an industrial research laboratory." *American Documentation* 11 (1) January 1960 : 9-17.

Markee, Katherine M. "Where is the library researcher?". In Irene Braden Hoadley and Alice S. Clark eds. *Quantitative Methods in Librarianship: Standards, Research, Management*—Proceedings and Papers of an Institute held at Ohio State University, 3-16 August 1969. Connecticut : Greenwood Press, 1972. 119-121.

Marquis, Donald G. and Thomas J. Allen. "Communication Patterns in applied technology." *American Psychologist* 21 (11). November 1966: 1052-1060.

Martin, Miles W. "Use of random alarm devices in studying scientists' reading behaviour." *IRE Transaction on Engineering Management* 9 (2) June 1962 : 66-71.

Martin, Miles W. and Russell L. Ackoff. "Dissemination and recorded scientific information." *Management Science* 9 1963 : 322-336

Martin Jr, Miles W. "Measurements of value of scientific information". In Burton V. Dean ed. Operations Research in Research and Development : Proceedings of a Conference at Case Institute of Technology.

New York : Robert E. E. Krieger Publishing Co., 1978.

Martyn, John. "Report of Investigation on Literature Searching by Research Scientists." London : Aslib, 1964.

— — "Uninternational duplication of research." New Scientist 21 (377) 6 February 1964 : 338.

— — "Information needs and uses". In Carlos A. Cuadra ed. *Annual Review of Information Science and Technology.* Washington, D.C. : ASIS, 1974. Vol. 9 : 3-23.

Maslow, Abraham H. "Psychology of Science : A Reconnaissance" South Bend : Regenry / Gateway, 1969.

McGarry, Kevin. *"Communication, Knowledge and the Librarian".* London: Clive Bingley, 1975.

Meadows, A. J. *"Communication in Science".* London : Butterworths, 1974.

Meadows, A.J. and J.G. O'Connor. Investigation of Information Sources and Information Retrieval in Astronomy and Space Sciences. OSTI 5044. Leicester : University of Leicester, Astronomy Department, 1969.

Meltzer, Leo. "Scientific productivity in organisational settings." *Journal of Social Issues* 12 1956 : 32.

Meltzer, Leo and James Slater. "Organisational structure and performance and job satisfaction of scientists." *American Sociological Review* 27 (3) January 1962 : 351-362.

Menzel, Herbert. "Planned and uplanned scientific communication". In *Proceedings of the International Conference on Scientific Information,* Washington, D.C., 16-21 November 1958. Washington, D.C.: National Academy of Sciences, National Research Council, 1959. 1: 199-243.

— — "Information needs of current scientific research." *Library Quarterly* 34 (1) January 1964 : 4-19.

— — "Information needs and uses". In Carlos A. Cuadra ed. *Annual Review of Information Science and Technology.* New York : Interscience Publishers for ADI, 1966. Vol. 1 : 41-69.

— — "Sociological perspectives on the information gathering practices of the scientific investigator and the medical practitioner". In D.Mc. Cord ed. *Bibliotheca Medica : Physician for Tomorrow,* Dedication of the Countway Library of Medicine. Boston : Harvard Medical School, 1966.

— — "Scientific communication : Five sociological themes," *American Psychologist* 21 (11) November 1966 : 999-1004.

— — "Informal communication in science : Its advantages and its formal analogues". In C. Rawski ed. *Toward a Theory of Librarianship-Papers in Honour of Jesse Shera.* New York : Syracuse University Press, 1966/1971.

— — Formal and Informal Satisfaction of Information Requirements of Chemists. PB-173261. Columbia : Columbia University, Bureau of Applied Social Research, 1970.

— — "Unplanned acquisition of information in the experience of polymer chemists." *IEEE Transactions on Professional Communication* PC-15 (2) June 1972 : 39-42, 47-50.

— — "Can science information needs be empirically determined?". In Thayer Lee. ed. *Communication : Concepts and Perspectives.* Washington, D.C. : Sparton Books, 1979, 279-294.

Merta, A. "Report on the results of experimental investigation into the users needs and requirements in the Czechoslovak Academy of Science". In *Second Symposium of Information Specialists of the Czechoslovak and Polich Academies of Sciences, Smolenice, Czechoslovakia,* October, 1967, Prague : Library of the Czechoslovakia Academy of Science, 1968.

Merton, Robert K. *"Social Theory and Social Structure"* New York : Free Press, 1957.

— — "Matthew effect in science". *Science* 159 (3810) 5 January 1968 : 56-63.

Meyer, Alan. "Some important findings in catalog use studies". In F.W. Lancaster ed. *The Measurement and Evaluation of Services.* Washington, D.C. : Information Resources Press, 1977. 69.

Meyer, Arnoud De. "Technological lifecycle approach to the organisational factors determining gatekeeper activities." *R&D Management* 14 (4) October 1984 : 239-246.

Michel, J. "Information needs and user studies". In EURIM : A European Conference on Research into the Management of Information Services and Libraries, 20-22 November 1973, UNESCO, Paris London: Aslib, 1974. 2-3.

Mick, Colin K. "Human factors in information work". In Alan R. Benenfeld and Edward John Kazlauskas eds. Communicating Information : *Proceedings of ASIS 1980 Meeting,* Ananheim, California, 5-10 October 1980. New York : ASIS, 1980. 17 : 21-23.

Mick, Colin K. et al. "Towards usable user studies." *Journal of the American Society for Information Science* 31 (5) September 1980 : 347-356.

Miller, G.A. "Magical number seven, plus or minus two : some limits on our capacity for processing information." *Psychological Review* 63 1956: 81-97.

Miwa, Makiko. "Creation and transfer of technological information at a research front." *Library and Information Science* (20) 1982 : 45-62.

Mooers, Calvin N. "Mooer's law or why some retrieval systems are used and others are not." *American Documentation* 11 (3) July 1960 : ii.

Moor, William Chattle. Empirical Study of the Relationship between Personality Traits of Research and Development Personnel and Dimensions of Information Systems and Sources. PhD Thesis. Illinois : Northwestern University, 1969.

——"Development and preliminary test of behaviour related dimensions of information systems." *Journal of the American Society for Information Science* 23 (1) January-February 1972 : 50-57.

Moss, L. "Industrial technologists and sources of information." In *The Direction of Research Establishments*. London : HMSO, 1957.

Mote, L.J.B. "Reasons for the variations in the information needs of scientists." *Journal of Documentation* 18 (4) December 1962 : 169 175.

Mulchenko, Z.M. et al. "On scientometrical characteristics on information activities of leading scientists." *Scientometrics* 1 (4) May 1979 : 307-325.

Mullins, Nicholas C. "Organisational approach to the informal communications among scientists". A Paper presented at Eastern Sociological Association Meeting, April 1967. N. p. n. p.

— — "Development of a scientific speciality : The phase group and the origins of molecular biology." Minerva 10 (1) 1972 : 51-82.

— — "Development of specialities in social science : the case of ethnomethodology." *Science Studies* 3 (3) 1973 : 245-273.

Myers, J.M. "Information as a saleable commodity". In Accelerating Innovation : Papers given at a Symposium held at the University of Notingham, March 1969. Organised by Aslib, U.K. Ministry of Technology and Research Associations. London : Aslib, 1970. 26-29.

Myers, Summer. Industrial Innovations : Their Characteristics and Their Scientific and Technical Information Bases : A Special Report to the National Science Foundation. Washington, D.C. : National Planning Association, 1966.

Myers, Summer and Donald G. Marquis. Successful Industrial Innovations: A Study of Factors Underlying Innovation in Selected Firms. NSF-69-17. Washington, D.C. : U.S. Government Printing Office, 1969.

Nagpaul, P. S. and S. Pruthi. "Problem-solving and idea generation in R&D: the role of informal communication" *R&D Management* 9 (3) July

1979 : 147-149.

National Research Council, Committee on Information in the Behavioural Sciences. Communication Systems and Resources in the Behavioural Sciences. Publication No.1575. Washington : National Academy of Sciences, 1967.

North American Aviation, Avionics Division. DoD User Needs Study, Phase II. Flow of Scientific and Technical Information within the Defense Industry. Final Technical Report. 3 vols. DoD C6-2442/030. California : North American Aviation, 1966.

O'Connor, Jean G. "Growth of multiple authorship". A Paper presented at 7th DRTC Seminar, 1969, Bangalore. Bangalore : DRTC, 1969. Paper JA, 463-486.

O'Gara, P.W. Physical Location as a Determinant of Communication Possibility among R&D Engineers. MS Thesis. Cambridge, Massachusetts : MIT, Alfred P.Sloan School of Management, 1968.

Oldman, Christine. "Value of Information." Library Research Occasional Paper No.1. Bradford : MCB Publications, 1976.

Olson, Edwin F "User population characteristics related to library utilisation". In Irene Braden Hoadley and Alice S. Clark eds. *Quantitative Methods in Librarianship : Standards, Research, Management* - Proceedings and Papers of an Institute held at the Ohio State University, 3-16 August 1969. Connecticut : Greenwood Press Inc., 1972. 108-112.

O'Reilly, Charles A. Superiors and peers as information sources and individual decision making performance., EDRSED-143949. 1977.

Orpen, Christopher. "Effect of managerial distribution of scientific and technical information on company performance." *R&D Management* 15 (4) October 1985 : 305-308.

Orr, Richard H. "Scientist as an information processor : A conceptual model illustrated with data on variables related to library utilization". In C.E. Nelson and D.K. Pollock eds. *Communication among Scientists and Engineers.* Massachusetts : D.C. Heath Lexington, 1970. 143-190.

Osburn, Charles B. "Non-use and loser studies in collection development." *Collection Management* 4 (1/2) Spring/Summer 1982 : 45-53.

Paisley, William J. Extent of Information Seeking as a function of Subjective Certainly and the Utility of the Information. PhD Thesis. Stanford University, California, 1965.

Paisley, William, J. "Flow of (Behavioural) Science Information : A Review of the Research Literature" Palo Alto ; Stanford University, Institute for Communication Research, 1965.

——"Information needs and uses". In Carlos A. Cuadra ed. *Annual Review of Information Science and Technology.* Chicago : Encyclopedia Brittanica Inc., 1968. Vol. 3 : 1-30.

Paisley, William J. and Edwin B. Parker. Scientific Information Exchange at an Interdisciplinary Behavioural Science Convention. PB-174837. Stanford, California : Stanford Institute for Communication Research, 1967.

Pao, Miranda Lee. "Coauthorship and productivity". In Alan R. Benenfeld and Edward John Kazlauskas eds. Communicating Information : *Proceedings of the 43rd ASIS Annual Meeting*, Anahein, California, 5-10 October 1980. New York : ASIS, 1980. 17 : 279-281.

Parker, Christopher E. "Use of external current awareness services at Southampton University." *Aslib Proceedings* 25 (1) January 1973 : 10-11.

Parker, Edwin B. and William J. Paisley. "Research for psychologists at the interface of the scientist and his information system." *American Psychologist* 21 (11) November 1966 : 1061-1071.

— — "Research for psychologists at the interface of the scientist and his information system". In Tefko Saracevic comp. & ed. *Introduction to Information Science.* New York : R.R. Bowker, 1970. 85-94.

Passman, Sidney. *Scientific and Technological Communication.* London : Pergamon Press, 1969.

Pelz, Donald C. and Frank M. Andrews. "Scientists in Organisations : Productive Climates for Research and Developments" Michigan : The University of Michigan, Institute for Social Research, 1976

Perrucci, Robert and Robert A. Rothanman "Obsolescence of knowledge and the professional career". In Robert Perrucci and Joel E. Gerstl eds. *The Engineers and the Social System.* New York : Wiley, 1969. 247-276.

Persson, O. "Critical comments on the gatekeeper concept in science and technology." *R&D Management* 11 (1) January 1981 : 37-40.

Pings, Vern M. and Fanny Anderson. Study of the Use of Wayne State University Medical Library, Pt.1 Detroit, Michigan : Wayne State University, School of Medicine Library and Biomedical Information Service Centre, 1965.

Preisler, Werner. "Methods of determining individual information needs." *Informatik* (1) 1971 : 40-49.

Prentice, Ann E. "Information seeking patterns of selected professionals." *Public Library Quarterly* 2 (2) Spring 1980 : 27-28, 58-60.

Price, Derek J. De Solla. *Science Since Babylon.* New York : Yale University

Press, 1961.

— — "Science of Science". In M. Goldsmith ed. Society and Science. New York : Simon & Schuster, 1965.

— — "Is technology historically independent of science?" : A study of statistical historiography." *Technology and Culture* 6 (4) Fall 1965: 553-568.

— — "Structures of publications in science and technology". In William H. Gruber and Donald. G. Marquis eds. Factors in Transfer of Technology. Massachusetts : MIT Press, 1969. 91-104.

— — "Citation measures of hard science, soft science, technology and nonscience". In Carnot E. Nelson and Donald K. Pollock eds. *Communication among Scientists and Engineers.* Lexington, Mas sachusetts, Heath Lexington Books, 1970. 3-22.

— — "Some remarks on elitism in information and the invisible college phenomenon in science." *Journal of the American Society for Information Science* 22 (2) March-April 1971 : 74-75.

Price, Derek J. De Solla and Donald Deb Beaver. "Collaboration in an invisible college." *American Psychologist* 21 (11) November 1966 : 1011-1018.

Pritchard, A. Information Transfer : Some Applications of Graph Theory and the Geography of Transport. London : The City of London Polytechnic, 1977.

Pruthi, S. and P. S. Nagpaul. "Communication patterns in small projects." *R&D Management* 8 (2) April 1978 : 53-58.

Pryor, Harold E. "Listening to the user : A case study". In problem of Optimization of User Benefit in Scientific and the Technological Information Transfer : Conference Proceedings of the Technical Specialists Meeting, 1975, October 8-9, Compenhagen, Denmark. Paris : AGARD, NATO, 1976. 11.1-11.7.

Putt, Archibald. *Putt's Law and the Successful Technocrat.* New York : Exposition Press, 1981.

Raising, Miles L. et al. "How biomedical investigators use library books." *Bulletin of the Medical Libraries Association* 54 April 1966 : 104-107.

Raitt, David Iain. Communication and Information Seeking and Use Habits of Scientists and Engineers in International Organisations and National Aerospace Research Establishments in Europe. PhD Thesis. Loughborough : Loughborough University of Technology, 1984.

— — "Information needs of scientists and engineers in aerospace". In Technical Information Panel Specialists Meeting on the Value of

Information as an Integral Part of Aerospace and Defence R&D Programmes, Cheltenham, U.K., 4-5 September 1985. AGARD-CP-385. Paris : AGARD, NATO, 1985. 3.1-3.5.

— — "Information seeking and communication habits off scientists and engineers". In Carol A. Parkhurst ed. *Proceedings of the 48th ASIS Annual Meeting,* Las Vegas, Nevada, 20-24 October 1985. New York: Knowledge Industry Publications for ASIS, 1985. 22 : 319-323.

Randall, Gordon E. "Literature obsolescence at a British and an American aeronautical library." *Special Libraries* 50 (9) November 1959 :447-450.

Rao, N.P. "Survey of Scientific Literature and Information Activities in India (R&D) and Abroad (Global Studies in Growth Pattern)." Delhi : Scientific Information Bureau, Defence Science Laboratory, 1966.

Rao, U.R. "Space Technology : Its Relevance to the Development of the nation." Bangalore : Bangalore University, 1976.

Rawdins, Eugene. "Field survey of information needs of industry sci/tech library users". *In Proceedings of the 38th ASIS Annual Meeting.* New York : Greenwood, 1975. 12 : 41-42.

Ritchie, E. "Communication networks : tools for the efficient management of R&D." *R & D Management* 7 (2) April 1977 : 85-88.

Robbins, J. C. "Communication among scientists and engineers : A commentary." *IEEE Transactions on Professional Communication* PC-15 (2) June 1972 : 65-67.

Roberts, Norman. "Draft definitions; information and library needs, wants, demands and uses : A comment." *Aslib proceedings* 27 (7) July 1975: 308-313.

Robertson, Andrew. "Behaviour patterns of scientists and engineers in information seeking for problem solving." Aslib Proceedings 26 (10) October 1974 : 384-390.

Robertson, Andrew and Roy Rothwell. Scientific and Technical Information Needs in R & D Laboratories : Reports of a Workshop, 19-20 February 1975. BLR&DR-5255. London : Polytechnic of Central London, 1975.

Rogers, Everett M. "Mass media and interpersonal communication". In Ithiel de Sola Pool et.al.eds. Handbook of Communication. Chicago: Rand McNally College Publishing Co., 1973. 290-310.

Rosenberg, Victor. Application of Psychometric Techniques to determine the Attitudes of Individuals towards Information Seeking. MS Thesis. AD 637 713; AFOSR 724-66. Bethlehem, Pa. : Lehigh University, 1966.

——"Factors affecting the preferences of industrial personnel for information gathering methods." *Information Storage and Retrieval* 3 (3) July 1967 : 119-127.

——"Factors affecting the preferences of industrial personnel for information gathering methods". In Tefko Saracevic comp. & ed. *Introduction to Information Science.* New York : R.R.Bowker, 1970. 95-100.

Rosenbloom, Richard S. et al. Technology Transfer and the Flow of Technical Information in a Large Industrial Corporation. 2 Vols. Cambridge, Massachusetts : Harvard University, Graduate School of Business Administration, 1965.

Rosenbloom, Richard S. and Francis W. Wolek. Technology, Information and Organisation : Information Transfer in Industrial R & D. Boston, Massachussets : Harvard University, 1967.

— — Technology and Information Transfer : A Survey of Practice in Industrial Organisations, Boston : Harvard University, Graduate School of Business Administration, 1970.

Rothwell, Roy. "Patterns of information flow during the innovation process." *Aslib Proceedings* 27 (5) May 1975 : 217-226.

— — "Information patterns in innovation". In Robertson, A. and Roy Rothwell. Scientific and Technical Information Needs in R & D Laboratories : Reports of a Workshop, 19-20 February 1975. London: Polytechnic of Central London, 1975.

Rothwell, Roy et al. "SAPPHO updated : Project SAPPHO Phase II." *Research Policy* 3 (3) 1974 : 258-291.

Rothwell, Roy and A. B. Robertson. "Contribution of poor communication to innovation failure." *Aslib Proceedings* 27 (10) October 1975 : 393-400.

Rouse, William B. "Note on the creativity inengineering : implications for supporting system design." Information Processing and Management 22 (4) 1986 : 279.

Rowley, J.E. and C.M.D. Turner. "Dissemination of Information." London: Andre Deutsch/a Grafton Book, 1978.

Rubenstein, Albert H. et al. "Some preliminary experiments and a model of information seeking style of researchers". In *Proceedings of 20th National Conference on the Administration of Research*, Miami, Florida, October 1966. N. p. n. p., 1966.

– "Explorations on the information seeking style of researches". in Carnot E. Nelson and D.K. Pollock eds. *Communications among Scientists and Engineers.* Massachusetts : Heath Lexington Books, 1970. 209-232.

Salasin, J. and T. Cedar. "Person to person communication in applied research service delivery setting." *Journal of the American Society for Information Science* 36 (2) March 1985 : 103-115.

Saracevic, Tefko. "Information needs of less developed countries". In Roy D. Tully and Ronald R. Deultgen eds. Information Choices and Policies : Proceedings of the 42nd ASIS Annual Meeting, Minneopolis, Minnesota, 14-18 October 1979. New York : Knowledge Industry Publications for ASIS, 1979. 16 : 91-98.

——"Perception of the needs for scientific and technical information in less developed countries." *Journal of Documentation* 36 (3) September 1980 : 214-267.

Sayer, J. "DoD user survey". In A.G. Hoshovsky ed. *Proceedings of the 1st USAF Scientific and Technical Information Conference*, 30 September - 4 October 1963, Dayton, Ohio. Washington : U.S. Department of Commerce, 1963. 44-50.

Scates, D.E. and A.V. Yeomans. Activities of Employed Scientists and Engineers for keeping Currently Informed in their Fields of Work. Washington, D.C. : American Council of Education, 1950.

Schermerhorn, John R. Jr. et al. "Managing Organisation Behaviour" 2nd ed. New York : John Wiley, 1982.

Schussel, George. "Advent of information and inquiry services." *Journal of Data Management* 7 (9) September 1969 : 24-31.

Scott, Christopher. "Science of science : What scientists read and why?" Discovery 20 (3) March 1959 : 110-114.

——"Use of technical literature by industrial technologists". In *Proceedings of the International Conference on Scientific Information,* Washington, D.C., 16-21 November 1958. Washington, D.C. : National Academy of Sciences, National Research Council, 1959. I: 245-266 (Preprints p. 235-256).

——Use of Technical Literature by Industrial Technologists : A Study of Technical Information in the Electrical Industry 1957-58 (Survey designed by Leslie T. Williams). The Social Survey Report No. 245. London : The Social Survey, 1960.

Sears, Roberts R. "Theoretical framework for personality and social behaviour." *American Psychologist* 6 (8) August 1951 : 475-483.

Shalini, R. and K. S. Chudamani. "Channel selection by Indian scientists for communication". In R. N. Sharma and S. Seetharama eds. *Proceedings of the Seminar on Primary Communications in Science and Technology in India,* 4-8 December 1978, Bangalore. New Delhi : CSIR, 1978. 103.

Shaw, Ralph R. Studies on the Use of Literature in Science and Technology. Washington, D.C. : National Science Foundation, n.d.

— — "Pilot Study of the Use of Scientific Literature by Scientists." Washington, D.C./Metuchen, N.J. : NSF/Scarecrow Reprint Corporation, 1956/1971.

Shekhurin, D.E. "Criterion of information needs." *Nauchnotekhnicheskaya Informatsiya, Seriya* 1 (5) 1968 : 3-7.

Shera, Jesse H. How Engineers Can Keep Abreast of Professional and Technical Development. ASME Paper No. 62-MD-5. New York : ASME.

Sherwin, C. W. amd R. S. Inemson. First interim Report on Project Hindsight (Summary). Washington, D.C. : Office of the Director of Defence Research and Engineering, 1966.

Sheth, Jagdish N. "Model of user behaviour for scientific and technical information". In William R King and Gerald Zaltman eds. *Marketing Scientific and Technical Information.* Colorado : Westview Press, 1979. 49-65.

Shockley, William. "On the statistics of individual variations of productivity in research laboratories." *Proceedings of the IRE* 45 (3) March 1957: 279-290.

Shotwell, T. K. "Information flow in an industrial laboratory - A case study." *IEEE Transactions on Engineering Management* EM-18 (I) February 1971 : 26-33.

Shuchman, Hedvah L. "Informal information networks and women in engineering." In Alan R. Benenfeld and Edward John Kazlaska eds. *Communicating Information : Proceedings of the 43rd ASIS Annual Meeting,* Anahein, California, 5-10 October 1980. New York : ASIS, 1980. 17 : 242-245.

— — Information Transfer in Engineering. Rep-461-46-27. Connecticut : Futures Group Inc., 1981.

— — "Information technology and the technologist : A report on a national study of American engineers." *International Forum on Information and Documentation* 7 (1) January 1982 : 3-8.

Sieber, H F. "Methodology of the DoD Scientific and technical information use study". In *Proceedings of the 27th Annual Meeting of the ADI on Parameters of Information Science,* Philadelphia, Pa., 5-8 October 1964. Washington, D.C. : ADI, 1964. (1) : 235-242.

Singh, B. N. "Information needs of engineering scientists." *International Library Review* 13 (2) April 1981 : 167-188.

— — "Scientists and social scientists as information users : A comparison of results of science user studies with the investigation into information requirements of the social sciences." *Journal of Librarianship* 5 (2) April 1973 : 138-156.

Slater, Margaret. "User and library surveys". In H.A.Whatley. ed. *British Librarianship and Information Science 1966-70.* London : The Library Association, 1972. 232-256.

— — "Non-Use of Library-Information Resources at the Workplace : A Comparative Survey of Users and Non-Users of Onsite Industrial Commerical Services."London : Aslib, 1984.

Slater, Margaret and P. Fisher. *Use Made of Technical Libraries.* Aslib Occasional Publication No.2. London : Aslib, 1969.

Slater, Margaret and Stella Keenan. "Methods of conducting a study of physicists requirements in current awareness in the US and the UK". In Proceedings of the American Documentation Institute. New York: ADI, 1967. 4 : 63-67.

— — Results of a Questionnaire Survey on Current Awareness Methods Used by Physicists Prior to Publication of Current Papers in Physics. INSPEC/1; AIP/CPP1. London/New York : IEE/AIP, 1967.

Spirit, Jiri and Ladislav Kobnovec. "Systematically ascertaining requirements of scientists for information". In *International Conference on Scientific Information,* Washington, D.C., 16-21 November 1958. Washington,D.C. : National Academy of Sciences, National Rsearch Council, 1959. I : 189-194.

Sridhar, M. S. "Study of correlation of use, citation and publishing of papers on Journals by Indian Space Technologists." A paper presented at XV All India IASLIC Conference, Bangalore, 26-29 December 1985, NPNP *(Collection Management,* 12 (112) 1990: 71-79)

— — "Mathematical approach to relations in thesauri." *Journal of Library and Information Science* 5 (1) June 1980 : 77-79.

— — "Study of library visits and inhouse use of library documents by Indian space technologists." *Journal of Library and Information Science* 7 (2) December 1982 : 146-158.

— — "User participation in collection building in a special library ; A case study." *IASLIC Bulletin* 28 (3) September 1983 : 99-106.

— — "Document reservation pattern in a special library : A case study." *Library Science with a slant to Documentation* 20 (1) March 1983 : 39-48.

— — "Use of technical reports and standards." *IASLIC Bulletin 29* (3) September 1984 : 99-106.

— — "A Study of co-authorship and collaborative research among Indian space technologists." *R&DManagement* 15 (3) July 1985 : 243-249.

——"Citing pattern of Indian space technologists." *International Library Review* 17 (3) July 1985 : 259-274.

——"A Case study of lentout use of a special library." *Library Science with a Slant to Documentation* 22 (1) March 1985 : 19-34.

—— "Book procurement delay : A demotivator to user participation in collection development". In Building Library Collections and National Policy for Library and Information Services : Seminar Papers of 30th All India Library Conference, Rajasthan University, Jaipur, 28-31 January 1985. Delhi : ILA, 1985. 329-334.

——"Pattern of card catalogue consultation in a special library." *IASLIC Bulletin* 31 (1) March 1986 : 9-16.

——"Subject and longitudinal use of books by Indian space technologists." *Collection Management* 8 (1) Spring 1986 : 101-115.

——"Use of current journals by Indian space technologists." *The Serials Librarian* 10 (3) Spring 1986 : 77-93.

——"Information and communication behaviour of women space technologists." *R&D Management* 17 (4) October 1987 : 301-309.

Srinivasan, S. "Impact of user study in the improvement of documentation services in developing countries". In Users of Documentation :Papers Presented at the 35th Conference and International Congress of Documentation, Buenos Aires City, 14-24 September 1970. Paris: FID, 1970. 1-14.

Strain, Paula M. "Engineering libraries : A user survey." *Library Journal* 98 (9) 1 May 1973 : 1446-1448.

Streatfield, David. "Senior manager's information needs." *Aslib Proceedings* 36 (11/12) November/December 1984 : 419-423.

Summers, Edward G. et al. "Effect of personal, professional and psychological attributes and information seeking behaviour on the use of information sources by educators." *Journal of the American Society for Information Science* 14 (1) January 1983 : 75-85.

Sutton, J. R. "Information requirements of engineering designers". In The Problems of Optimisation of User Benefit in Scientific and Technological Information Transfer : Panel Specilists' Meeting, Copenhagen, Denmark, 8-9 October 1975. Paris : AGARD, 1976. 12.1-12.8.

Swinburne, J. K. *Sources of Information Used by British and French Scientists : A Study of Two Groups.* MS Thesis. London : City University, 1981.

——"Information use and transfer by British and French scientists : a study of two groups." *Journal of Information Science* 6 (2/3) April 1983: 75-80.

Szwalbe, Jerzy. "Wybrane zagadnienia metodologii badania potrzeb uzytkownikow inte w swiette litertney zagranicznej (Selected problems in methodology of users needs :investigation in the light of foreign literature)." *Przeglad Biblioteczny* 48 (1) 1980 : 31-39.

Tagliacozzo, R. et al. "Access and recogition : from user's data to catalogue entries." *Journal of Documentation* 26 (3) September 1970 : 230-249.

Taylor, Robert L. "Technological gatekeeper." *R&D Management* 5 (3) 1975 : 239-242.

Taylor, Robert L. and James M. Utterback. "Longitudinal study of communication in research : Technical and managerial influences." *IEEE Transactions on Engineering Management* EM-22 (2) May 1975 : 80-87.

— — Question-Negotiation and Information Seeking in Libraries. AD-659168. Bethleham, Pa. : Lehigh University Centre for the Information Sciences, 1967.

Taylor, Robert S. "Question-negotiation and information seeking in libraries." *College and Research Libraries* 29 (3) May 1968 : 178-194.

Thomas, C. C. International Study of the Dissemination of Technical Information to Industry. The Social Survey Report No. 254. London: The Social Survey, 1954.

Thompson, James. *End of Libraries.* London : Clive Bingley, 1982.

Thorne, R. G. Survey of the Reading Habits of Scientific and Technical Staff at the RAE. England : Royal Aircraft Establishment, 1954.

Thorpe, W. V. "International statement on information exchange groups." *Science* 155 (3766) 10 March 1967 : 1195-1196.

Tomlin, B. "Inter-location technical communication in a geographically dispersed research organisation." *R&D Management* 11 (1) 1981 : 19-23.

Tornudd, Elin. Professional Reading Habits of Scientists Engaged in Research as Revealed by an Analysis of 130 Questionnaires. Master's Dissertation. Carnegie : Carnegie Institute of Technology, Carnegie Library School, 1953.

— — "Study on the use of scientific literature and reference services by Scandinavian scientists and engineers engaged in research and development". In Proceedings of the International Conference on Scientific Information, Washington, 16-21 November 1958. Washington, D.C. : National Academy of Sciences, National Research Council, 1959. 1 : 9-65, Preprint 1.

Trueswell, Richard L. "Some behavioural patterns of library users : the 80/20 rule." *Wilson Library Bulletin*, 43 (5) January 1969 : 458-461.

Trueswell, Richard W. "Article use and its relationship to individual user satisfaction." College and Research Libraries 31 (4) July 1970: 239-245.

Trueswell, Richard W. et al. Experiment in Measuring Certain Information Seeking Behaviour of X-Ray Crystallographers. Technical Report No.1. Illinois : Northwestern University, 1965.

Turner, W. A. "Information requirements of electronic research." *Aslib Proceedings* 12 (5) May 1960 : 186-190. U.S.Department of Defense. DoD User Needs Study. 2 Vols. Defense Docmentation Centre, 1965.

Unesco. *Unesco Statistical Year Book 1983*. Paris : UNESCO, 1983.

University of Cambridge, Library management Research Unit. "Factors affecting the use of seats in academic libraries." *Journal of Librarianship* 7 (4) October 1975 : 262-287.

University of Chicago, Graduate Library School. Requirements Study for Future Catalogs. Progress Report No.2. Chicago, Illinois : University of Chicago, 1968.

University of Sussex, Science Policy Research Institute. "Success and Failure in Industrial Innovation" London : Centre for the Study of Industrial Innovation, 1972.

Urquhart, D. J. "Distribution and use of scientific and technical information". In Report of the Royal Society Scientific Information Conference, 21 June-2 July 1948. London : The Royal Society, 1948. 408-419.

——"Distribution and use of scientific and technical information." *Journal of Documentation* 31 1948 : 222-231.

—— "Physics abstracting - use and users." *Journal of Documentation* 21 (2) June 1965 : 113-121.

Utterback, James M. Process of Technological Innovation in Industrial Firms. Ph.D Thesis. Cambridge, Massachusetts : MIT, Alfred P. Sloan School of Management, 1969.

Vagianos, Louis. "Information patterns of chemists in a university environment." *Journal of Chemical Documentation* 11 (2) May 1971 : 85-89.

Van Cott, Harold P. and Robert G. A. Kincade. Feasibility Study for Determining Requirements of Bibliographical Information Services and Systems. Final Report AIR-F-57-11/67-TR-7. Washington, D.C. : American Institute for Research, 1967.

Vickery, B. C. "Use of scientific literature." *The Library Association Record*

63 (8) August 1961 : 263-269.

— — "Information Systems" London : Butterworths, 1973.

Vincent, Harry F. and Abbas Mirakhor. "Relation between productivity, satisfaction, ability, age and salary in a military R&D organisation." *IEEE Transactions on Engineering Management EM*-19 (2) May 1972 : 45-53.

Voigt, Melvin J. "Researcher and his sources of scientific information." Libri 9 (3) 1959 : 177-193.

— — "Scientists' Approaches to Information" Chicago :ALA, 1961.

Waldhart, T. J. Relationship between the Citation of Scientific Literature and the Institutional Affiliation of Engineers. Ph.D Thesis. Bloomington, Indiana : Indiana University, 1973.

Waldhart, T. J. and E. S. Waldhart. "Communication Research in Library and Information Science : Bibliography on Communication in the Sciences, Social Sciences and Technology." Littleton : Littleton Co., Libraries Unlimited, 1975.

Walsh, V. M. and A. G. Baker. "Project management and communication pattern in industrial research." *R&D Management* 2 (2) 1971 : 103-109.

Webb, J. and Donald T. Campbell. "Experiments on communication effects". In Ithiel de Sola Pool et. al. eds. *Handbook of Communication.* Chicago : Rand McNally College Publishing Company, 1973. 938-952.

Weick, Karl E. "Twigging of overload". In Harold B. Pepinsky ed. *People and Information.* New York : Pergamon, 1970. 67-129.

Weinberg, Alvin. "Science, government and information". In Manfred Kochen ed. *The Growth of Knowledge : Readings on Organisation and Retrieval of Information.* New York : John Wiley, 1967. 36-59.

Wersig, Gernot. "Communication theory and user analysis : The communication theory frame of reference". In Users of Documentation : Papers presented at the 35th Conference and International Congress of Documentation, Buenos Aires city, 14-24 September 1970. Paris: FID, 1970. 1-13.

— — "Zur systematik der denutzerborschung." *Nachnichten fur Dokumentation* 24 (1) February 1973.

Wersig, Gernot and Gunther Windel. "Potential basic research on users for new orientation of information services." *International Forum on Information and Documentation* 6 (4) October 1981 : 21-23.

Wilkin, Anne. "Some comments on the information broker and the technological gatekeeper." *Aslib Proceedings* 26 (12) December

1974 : 477-484.

— — Information Needs of Practitioners : A Review of the Literature. BLR&DR 5611. London : The London University, 1981.

Willcoock, H. D. Dissemination of Technical Information in Industry. The Social Survey Report No. 193. London : The Social Survey, 1953.

Wilson, Edward C. "Information discrimination : A human habit." *Canadian Journal of Information Science* 1 (1) 1976 : 59-64.

Wilson, Patrick. *Public Knowledge and Private Ignorance : Toward a Library and Information Policy.* London : Greenwood Press, 1977.

Wilson, T. D. "On user studies and information needs." *Journal of Documentation* 37 (1) March 1981 : 3-15.

— — "Cognitive approach to information seeking behaviour and information use." *Social Sciences Information Studies* 4 (2 & 3) April/July 1984 : 197-204.

Wilson, T. D. and D. R. Streatfield. "Structural observation in the investigation of information needs." *Social Sciences Information Studies* 1 (3) April 1981 : 173-184.

Woelfle, Robert M. "Role of the engineer in improving the communication of technical information." *IEEE Transactions on Professional Communication PC*-22 (1) March 1979 : 24-26.

Wolek, Francis W. "Engineer : His work and needs for information". In J.B.North ed. *Proceedings of the 32nd Annual ASIS Meeting.* New York : Greenwood, 1969.

— — "Complexity of messages in science and engineering : an influence on patterns of communication". In Carnot E. Nelson and Donald K. Pollock eds. *Communication among Scientists and Engineers.* Massachusetts : D.C. Heath Lexington, 1970. 233-265.

— — "Preparation for interpersonal communication." *Journal of the American Society for Information Science* 23 (1) January-February 1972 : 3-10.

— — "Managers and the distribution of scientific and technical information." *R&D Management* 14 (4) October 1984 : 225-228.

Wood, D. N. "Where mechanical engineers get their technical information." *The Chartered Mechanical Engineer* 14 (5) May 1967 : 212-215,220.

— — "Discovering the user and his information needs." *Aslib Proceedings* 21 (7) July 1969 : 262-270.

— — "User studies : A review of the literature from 1966-1970." *Aslib Proceedings* 23 (1) January 1971 : 11-23.

Wood, D. N. and D. R. L. Hamilton. Information Requirements of

Mechanical Engineers : Report of a Recent Survey. London : The Library Association, 1967.

Wood, J. and J. Ronayne. Science and Technology News Periodicals, 2 vols. OSTI-5142. England : University of Manchester, 1972.

Wooster, Harold. "Zoo and jungle - a comparison of the information practices of intelligence analysts and scientists". In Donald E. Walker ed. *Information System Science and Technology Washington* : Thompson, 1967. 307-315.

World Federation of Engineering Organisations, Committee on Engineering Information. Engineer's Needs for Scientific and Technical Information. By A.David et.al. Paris : WEFO, 1979.

Wysocki, A. "Studies of User's Needs in Scientific Information" Paris : International Council of Scientific Unions, 1968.

Yates, Bryan. "Pilkington technical communications system : a formalisation of the role of the technological gate-keepers." *Aslib Proceedings* 22 (10) October 1970 : 507-510.

Zaltman, Gerald. Scientific Recognition and Communication Behaviour in High Energy Physics. Ph.D Thesis. PB-179890. The Johns Hopkins University, Baltimore, Maryland. New York : AIP, 1968.

Zuckerman, Harriet. "Nobel laureates in science : patterns of productivity, collaboration and authorship." *American Sicological Review.* 32 (3) January 1967 : 391-403.

INDEX

CICIL LIST

CONCEPTS IN COMMUNICATION, INFORMATICS & LIBRARIANSHIP

(CICIL)

General Editor: **S.P. AGRAWAL**

CICIL-1 Development of Library Services in India: Social Science Information
S.P. Agrawal Rs. 150

CICIL-2 Development of Documentation in India: Social Science Information
S.P. Agrawal Rs. 250

CICIL-3 UNESCO and Social Sciences: Retrospect and Prospect
S.P. Agrawal and **J.C. Aggarwal** Rs. 250

CICIL-4 INDIANA : A Bibliography of Bibliographical Sources
M.K. Jain Rs. 200

CICIL-5 Dimensions of Library and Information Science: Kaula Festschrift
V. Venkatappaiah (Eds.) Rs. 600

CICIL-6 Book Numbers: Some Indian Methods
M.P. Satija and **S.P. Agrawal** Rs. 100

CICIL-7 National Policy for University Libraries in India: Problems and Perspectives
N.B. Inamdar and **L.S. Ramaiah** (Ed.) Rs. 250

CICIL-8 Media Utilisation for the Development of Women and Children
B.S. Thakur and **Binod C. Agrawal** Rs. 150

CICIL-9 Terrorism: An Annotated Bibliography
Susheela Bhan (Ed.) Rs. 275

CICIL-10 Transnational Library Relations: The Indo-American Experience
M.B. Konnur Rs. 175

CICIL-11 Geography of Marketing and Commercial Activities in India (Documentation on Research Information)
R.S. Dixit Rs. 130

CICIL-12 Information India 1988: Global View
S.P. Agrawal and **J.C. Aggarwal** Rs. 200

CICIL-13 Women's Education in India: Historical Review, Present Status and Perspective Plan
S.P. Agrawal and **J.C. Aggarwal** Rs. 360

CICIL-14 Cultural Communication in India: Role and Impact of Phonograms
J.M. Ojha Rs. 220

CICIL-15 Second Historical Survey of Educational Development in India: Select Documents 1985-1989
S.P. Agrawal and **J.C. Aggarwal** Rs. 490

CICIL-16 Documentation Encyclopaedia of UNESCO and Education
J.C. Aggarwal and **S.P. Agrawal** Rs. 800 (set of 2 parts)

CICIL-17 Attitude Towards Information: A Study of Post-Graduate Teachers and Researchers in Social Sciences
N.G. Satish (in press)

CICIL-18 Librarianship and Bureaucratic Organisation: A Study in the Sociology of Library Profession in India
P.K. Jayaswal Rs. 175

CICIL-19 Literature Search
K.A. Isaac Rs. 150

CICIL-20 Child Education in India: Index to Scholarly Writings in Indian Educational Journals and Newspapers since independence
S.P. Agrawal and **Naresh Kanta** Rs. 100

CICIL-21 Lok Sabha and Vidhan Sabha Elections in India 1989 & 1990: Process and Results with Comparative Study of Manifestos
S.P. Agrawal and **J.C. Aggarwal** Rs. 250

CICIL-22 Islamic Studies in India: A Survey of Human, Institutional and Documentary Sources
Mohamed Taher Rs. 180

CICIL-23 Panchayati Raj: An Annotated Bibliography
N.B. Inamdar, N.G. Satish and **Anil Takalkar** 290

CICIL-24 Information India 1989-90: Global View
S.P. Agrawal and **J.C. Aggarwal** Rs. 500

CICIL-25 First Handbook of Psychological and Social Instruments
Udai Pareek and **T. Venkateswara Rao** Rs. 500

CICIL-26 Educational and Social Uplift of Backward Classes: At What Cost and How? Mandal Commission and After
S.P. Agrawal and **J.C. Aggarwal** Rs. 275

CICIL-27 Mass Media and New Horizons: Impact of TV & Video an Urban Milieu
H.R. Trivedi Rs. 180

CICIL-28 Indian Writings on Education : An Indicator to Indian Educational Journals Grouped by Subject Deseriptors 1979 to 1986 (in two parts)
S.P. Agarwal Rs. 900 (per set)

CICIL-29 Nation in Crisis
S.P. Agrawal and **J.C. Aggarwal** Rs. 200

CICIL-30 Information India 1990-91: Global View
S.P. Agrawal and **J.C. Aggarwal** Rs. 500

CICIL-31 Development of University Libraries in India after Independence
O.P. Gupta Rs. 270

CICIL-32 Handbook on Child : With Historical Background
Pramila Pandit Barooah Rs. 320

CICIL-33 Weaker Section in Indian Society : Undiscovered Past, Uncertain Present and Unpredictable Future : Index to Writings in Indian Scholarly Journals and 10 National Newspapers Classi-

fied under 1500 Descriptors 1886-1990
S.P. Agrawal and **Rajeev Kumar Sharma** Rs. 600

CICIL-34 Communication Informatics and Librarianship in India : An Introductory Guide to Publications Based on a Survey of Libraries in Capital City of India
S.P. Agrawal and **Pushpa Rani Sharma** Rs. 720

CICIL-35 Problems of Collection Development in Special Libraries
M.S. Sridhar Rs. 225

CICIL-36 National Information Resources for Social Sciences in India : Mobilisation, Dispersal and International Enhancement
S.P. Agrawal Rs. 225

CICIL-37 Modern History of Punjab : A Look Back into Ancient Peaceful Punjab Focusing Confrontation and Failures Leading to Present Punjab Problem, and a Peep Ahead : Relevant Select Documents
J.C. Aggarwal and **S.P. Agrawal** Rs. 400

CICIL-38 India Information Sources : Economics and Business
Kyllikki Ruokonen and Manohar Lal Rs.

CICIL-39 Educational Planning in India : With a Slant to Educational Financing and Administration, Vol. 1, Reports of Committees and Commissions, Five Year Plans, Statistical Tables
J.C. Aggarwal and **S.P. Agrawal** Rs. 650

CICIL-40 Integrated Library System : Two Case Studies : Latin America and India
Maria Luiza Rigo Pasquarelli, Rosaly Favero Krzyanowski, Ines Maria M. Imperatriz and **S.P. Agrawal** Rs. 160

CICIL-41 Social Science Information in India: Effort towards Bibliographical Control
S.D. Vyas Rs. 300

CICIL-42 Higher Education in India Since Independence: U.G.C. and Its Approach
O.P. Gupta Rs. 360

CICIL-43 Government and Politics in India: A Bibliographical Study of Contemporary Scenario Chronicling Rajiv Gandhi Era
S.P. Agrawal and **Rajeev Kumar Sharma** Rs. 500

CICIL-44 Second Handbook of Psychological and Social Instruments
D.M. Pestonjee Rs. 300

CICIL-45 Development/Digression Diary of India; 3D Companion Volume to Information India: 1990-91
S.P. Agrawal Rs. 200

CICIL-46 Development/Digression Diary of India; 3D Companion Volume to Information India: 1991-92
S.P. Agrawal Rs. 300

CICIL-47 Information India 1991-92: Global View
S.P. Agrawal and **J.C. Aggarwal** Rs. 500

CICIL-48 Committees and Commissions in India 1974, A Concept's Project, Compiled by a team of professionals guided by Virender

Kumar (Vol. 12) Rs. 450

CICIL-49 Committees and Commissions in India 1975, A Concept's Project, Compiled by a team of professionals guided by Virender Kumar (Vol. 13) Rs. 450

CICIL-50 Committees and Commissions in India 1976, A Concept's Project, Compiled by a team of professionals guided Virender Kumar (Vol. 14) Rs. 450

CICIL-51 Committees and Commissions in India 1977, A Concept's Project, Compiled by a team of professionals guided by Virender Kumar (Vol. 15) (in four parts) Rs. 450 each

CICIL-52 Committees and Commissions in India 1978, A Concept's Project, Compiled by a team of professionals guided by Virender Kumar (Vol. 16) (in two parts) Rs. 450 each

CICIL-53 Subject Guide to Communication, Informatics and Librarianship in India (Based on a Survey of Library Holdings in Capital City of India)
S.P. Agrawal and **Pushpa Rani Sharma** Rs. 400

CICIL-54 Restructurring of Library and Information Science Curriculum
A. Tejomurty Rs. 250

CICIL-55 Committees and Commissions in India 1979, A Concept's Project, Compiled by a team of professionals guided by Virender Kumar (Vol. 17) (in two parts) Rs. 450 each

CICIL-56 Committees and Commissions in India 1980, A Concept's Project, Compiled by a team of professionals guided by Virender Kumar (Vol. 18) (in two parts) Rs. 450 each

CICIL-57 Third Historical Survey of Educational Development in India
S.P. Agrawal and **J.C. Aggarwal** Rs. 700

CICIL-58 Model Library Legislation
Velaga Venkatappaiah Rs. 400

CICIL-59 Indian Library and Information Science Literature (1990-91)
Sewa Singh Rs. 525

CICIL-60 Librarianship and Library Science in India
Mohamed Taher and **Donald Gordon Davis, Jr**. Rs. 350

CICIL-61 Comprehensive Glossary for Library & Information Science by
O.P. Gupta Rs.300

CICIL-62 Information India 1992-93:
S.P. Agrawal and **J. C. Aggarwal** Rs. 500

CICIL-63 Automation of Libraries and Information Science
R.K. Shukla Rs. 250

CICIL-64 Modern History of Jammu and Kashmir
J.C. Aggarwal and **S.P. Agrawal** Rs. 1500 (per set of 2 vols.)

CICIL-65 Library and Information Science: Emerging Trends:
Sister M. Esperanza, A. C. Rs. 250

CICIL-66 Mental Retardation in India (A Bibliography)
S. Venkatesan and **V.G.D. Vepuri** Rs. 275

CICIL-67 Information Management in Health Science Libraries in India
R.P. Dixit (*Foreword* **Dr. Uton Muchtar Rafei**) Rs. 550

CICIL-68 Information Behaviour of Scientists and Engineers:
M.S. Sridhar Rs. 500